Ravens, Crows, Magpies, and Jays

Ravens

Ravens
Crows
Magpies
and
Jays

TONY ANGELL

with a Foreword by J. F. Lansdowne

UNIVERSITY OF WASHINGTON PRESS

Seattle and London

Raven flying

ENDSHEETS: One hundred Mexican crows

Library of Congress Cataloging in Publication Data

Angell, Tony.
  Ravens, crows, magpies, and jays.

  Bibliography: p.
  1. Corvidae. 2. Birds—United States. I. Title.
QL696.P2367A54   598.8'64'0973   77–15185
ISBN 0-295-95589-9

Composition, Printing, and Binding: Kingsport Press
Paper: Mohawk Superfine
Design: Audrey Meyer

For Frank, Florence, Lee, and Nat

Raven chasing goshawk

Crow in flight over dunes near Tillamook Head, Oregon

Foreword

The crows and their kin are extraordinarily interesting birds, members of what may be the most intelligent avian family. The language and behavior of the different species are elaborate and the complexities of their lives form an absorbing subject for study.

As everyone is aware, some species have evil reputations, and for ages these birds have engaged the attention of men. By strong character and dark cunning the family as a whole has won, if not affection, then a reluctant respect and a place in the mythology of many races. The raven in particular has penetrated deeply into human consciousness and whenever it occurs, as here on America's Northwest Coast, it has been woven inextricably into the fabric of man's beliefs.

This inadvertent involvement with the cultures and affairs of people is only one of many fascinating matters touched upon in *Ravens, Crows, Magpies, and Jays;* together with the natural history and behavior of each species, all aspects of the birds' lives are fully explored in an easily comprehended way. Tony Angell's interest in birds and other animals began in the hills of his native California; later in the Pacific Northwest it matured into a deep commitment to nature and the environment as a whole. His large physical size and expansive manner do not conceal his innate sensitivity nor do they hide the concern for living things that is a guiding force in his life. There runs through the book a strong personal theme, for Angell's encounters and friendships with animals are used effectively to enliven the text. Perhaps unconsciously he pays a debt to the late Francis Lee Jaques and Florence Jaques, the well-known artist-naturalist couple who in the latter years of their lives became both friends and inspiration.

As an artist, Tony Angell commands increasing respect, and in *Ravens, Crows, Magpies, and Jays* he has used his abilities to produce a host of vigorous and illuminating drawings. These are not intended to be portraits but rather are expressions of the birds' personalities, depicting characteristic actions and attitudes. Attractive in themselves, the drawings are adjuncts to the writing.

The illustration medium of black-and-white is well chosen. Its economy and starkness are striking and very well express the artist's gifts. Angell's drawings owe little to the influence of others and in this originality and freedom lies their chief strength. Whenever an illustration is needed, one appears, and in it are seen the birds displaying, mobbing an owl, feeding, or simply disporting themselves. Nothing more is required.

J. Fenwick Lansdowne

Raven flying at River's Inlet with totem in background

Preface

An artist can choose to clarify or amplify those elements in his subject that are most fundamental, enduring, and essential. Over the centuries, birds have inspired painters, poets, and sculptors to treat their strength, mystery, elegance, and beauty. The birds of the family Corvidae possess all of these elements, with the addition of a subtle cerebral quality, challenging portrayal, which we infer from the respect accorded some corvids by human cultures and which we sense when we consider their social behavior. The proof of this quality becomes more apparent when we examine the birds' capacities as communicators, tool users, and problem solvers. To the degree that corvids do these things they are set apart from other avian families and it appears that no other birds approach their breadth of intelligence. As a familiar biological associate, particularly as a competitor for available food supplies, these birds provide insights into the struggle for survival that is common to all life. It is my intention in this book then to describe, in words and drawings, the uniqueness of the corvid family and to pay tribute to the beauty of their form, their personalities, and their endurability in a world that continues to lose its natural inheritance.

The 18 corvids (of the 116 total world species) occurring within the United States are representative of the whole family. They capably exploit different habitats, come in a variety of sizes, and range in coloration from drab to dazzling. As I found that corvid behavior could best be described under general headings, I decided first to introduce each species briefly and then to discuss the birds' activities in more detail in chapters treating relationships with humankind, social strategies, intelligence, and the like. The species descriptions in "The Cast of Characters" include basic information on measurements, coloration, range, and food habits. Unfortunately, for a few species some of this information was unavailable; these omissions will be corrected in subsequent printings of the book if the data become available.

In writing this book I have drawn both on personal experiences and on the field and laboratory studies of others who have also found corvids rewarding subjects of study. The Bibliography reflects the widespread interest that North American corvids have excited; while it is not the last word on the literature, it does provide the reader with the means to pursue more specific study of the family.

While my writings and drawings permit me to share adventures with others they also help me retain the original pleasure I had in distant places and at other times. These are often recollections of places that no longer exist as I knew them. The dusty green arroyos of my Southern California youth are long since suffocated beneath platforms of fill, and the hillsides are terraces of rooftops. Gone too are many of those grand redoubts of the northern raven that still remembered the passage of the long canoes, or the tread of people who worshiped the earth.

This composition has helped me reach more deeply into meanings of experiences shared with fellow life forms. Perhaps this sharing will open up avenues for discovery and adventure in the reader. Through these pursuits we may personally begin to fashion a formula for human behavior that again includes reverence for other life.

Dr. Gordon Orians, Director of the Institute of Environmental Studies of the University of Washington, generously provided the insight and criticism essential to bringing this manuscript to completion. I am very grateful for the considerable amount of time and energy he spent on my behalf. Likewise Professor Russell Balda gave willingly of his wisdom and provided no small amount of inspiration to the writer. I wish also to thank Peter Gellately, head of the Serials Division of the University of Washington, who was my unfailing friend in my research efforts. Dr. Gordon Alcorn helpfully provided me with a wide range of study specimens when needed and Mr. John Arvin, of Edinburgh, Texas, made an exceptional effort in providing me with information gathered from years of experience with particular southwestern Texas corvids.

Over the several years of the book's preparation the following individuals all in some kind and thoughtful way supported the fruition of my enterprise: Paul Banko, Daphne Begg, Dr. Marjorie Halpin, Dr. and Mrs. Bert Bender, Dr. and Mrs. Fred Harvey, David and Nan Munsell, Dr. and Mrs. Frank Richardson, Bill Holm, Ruth Kirk, Dr. Sievert Rohwer, Don Eckelberry, Fen Lansdowne, and of course my wife Noel, whose patience, sacrifice, and support have been the ingredients necessary for the completion of both drawings and text.

Tony Angell
September 1977

Contents

North American Members of the Family Corvidae

SUBFAMILY	GENUS	SPECIES
Garrulinae	*Perisoreus*	Gray jay
	Cyanocitta	Blue jay
		Steller's jay
	Aphelocoma	Scrub jay
		Mexican Jay
	Cyanocorax	Green jay
	Cissilopha	San Blas jay
	Psilorhinus	Brown jay
	Pica	Black-billed magpie
		Yellow-billed magpie
Corvinae	*Corvus*	Common raven
		White-necked raven
		Common crow
		Northwestern crow*
		Fish crow
		Mexican crow
		Hawaiian crow
	Gymnorhinus	Piñon jay
	Nucifraga	Clark's nutcracker

* Generally considered a race of the common crow.

The Cast of Characters

Gray jays eye a picnic ground

Gray Jay

Perisoreus canadensis

Length	10 in.	254.0 mm.
Wingspread	15 in.	381.0 mm.
Weight	2½ oz.	70.87 gm.

One summer I led an expedition of teachers and students on a two-week field journey that was to include the expanse of our Northwest from open ocean to the Columbia River Plateau. For the first several days we camped on the coast and shivered under the incessant rains. The students' romance with wilderness was fast dissolving. Near the end of the week, spirits low, we traveled inland to the Olympic Mountains. The snow was deep along Hurricane Ridge and our sullen line trudged along, showing little interest in discussing alpine ecology. The youngsters held fast to their city-bred skepticism when I spoke of a bird so fearless that, if you had food to offer, it would fly from the trees to your hand. As the mountain gods would have it, there were gray jays in the vicinity, and, when I took a cracker from my pack to prove the point, one of the birds winged out of the clouds and alighted on my hand. His appearance was the magic our trip needed, and I like to remember that, as he flew off before astonished faces, the clouds began to shred, the breeze warmed, and a blue sky opened up over us. The rest of our journey was bright and dry.

These are birds of the high plateaus and mountains of the West and the northern coniferous forests of the East. Their populations follow the Rockies, Sierra Nevada, and associated ranges south into California, Colorado, and New Mexico.

There are two distinctive adult color patterns: a darker one, characteristic of northern birds (*P.c. canadensis*), and a lighter one, typical in the Rocky Mountains (*P.c. capitalis*). The fledgling plumage of the birds is also distinctive, for, unlike the adults with their shades of light grays and whites, the youngsters are uniformly charcoal gray.

Nesting as early as February, the female incubates from three to five eggs that are gray or greenish-white and touched with spots of olive-buff. Their nests are exceptionally well insulated and perfectly shaped to fit the contours of the incubating parent.

Although the gray jay is well known for eating a wide range of camp fare, including chunks of soap, bacon, and biscuits, its usual diet consists of assorted insects of coniferous forests. The jay will search the undersides of branches for egg cases and later catch the emergent insects like a flycatcher. What isn't eaten immediately is cached in some secret hollow or as a "bolus" in the needles of fir trees. These caches sustain the birds through the winter months and provide food for their young, which hatch early in the year.

Blue jay "anting"

Blue Jay

Cyanocitta cristata

Length	11¼ in.	285.73 mm.
Wingspread	16¼ in.	412.75 mm.
Weight	3¼ oz.	92.1375 gm.

As a child in Los Angeles, I watched the mindless momentum of
bulldozers fill stream beds and crush oak forests, fulfilling
the dreams of a freeway and parking lot planner. With obscene
indifference, the homes of my wild associates were destroyed,
and then they too were gone. I doubt that those faint memories of
nature would have sustained my interest, and, had it not been
for boyhood summers in the Michigan woods, I could very
likely have settled into a conditioned somnolence of urban
life.

Fortunately, my grandparents and their parents before them
lived at the edge of a tiny town, and more in the flow of woodland
life than the action of the human community. I started each
summer day with a hike into the oak and beech forests. Stretching
out to look into the canopy of trees, I soon found the blue jays
that chased, scolded, courted, mobbed, and serenaded me
with whisper songs. Their animated lives opened up all the
possibilities of avian behavior and beauty, and surely set the
course of my life's interest.

Probably our best-known jay, the blue jay is common from the
eastern flanks of the Rocky Mountains to the Atlantic Coast. Over
most of its range it is the only jay, but it overlaps with gray
jays in the north, scrub jays in Florida, and occasionally occurs in
the range of Steller's jays. There are four races, the southeastern
birds being slightly smaller than their northern counterparts.

Their four to six eggs come with a variety of ground colors from
cream to olive, and pea green to bluish- or gray-green, typically
spotted with various browns and lavender. In the spring blue jays
take the opportunity to feed on the eggs and offspring of other
birds, but for the balance of the year they are voracious seed
and insect eaters. In winter months they will, in the corvid
tradition, feed from caches of seeds and nuts.

Blue jays live about six years or less in the wild. An exception
to this was one wild jay that reached an age of fourteen and
one-half years (Kennard 1975). A captive blue jay in Ontario,
Canada, was doing well eighteen and one-half years after
its capture (Judd 1975).

Steller's jays

Steller's Jay

Cyanocitta stelleri

Length	12 in.	304.8 mm.
Wingspread	17½ in.	444.5 mm.
Weight	3 oz.	84.75 gm.

We live where there are still forests so ancient it is difficult to conceive of the time of their youth. One spring we rafted through the heart of such a place, following the courses of the Ozette River. Along the way, Steller's jays came to witness our passage, and it was they who gave me some sense of time. I imagined their ancestors, some sixty generations before, following the courses of long canoes that may have ventured inland from the coast. I wondered if, locked within some corvid code, were chronicles of the human adventures they had witnessed, and if their brethren of the Far North had bothered to note the scientist-explorer who had named them.

This is a jay of the West from the coastal reaches of Alaska, down through western Canada, south into northern Baja California. They also occur in the mountains of Mexico and Central America, as far south as pine forests occur. They as a rule enjoy coniferous forests, both coastal and mountain. In dry portions of their range they inhabit mixed woodlands of pine, oak, and juniper.

Steller's jays prefer to nest in conifers. The female lays three to five greenish-blue eggs, spotted with brown, and when she is incubating, in late May, the usually boisterous pair are quiet and retiring. I have often watched a nesting pair of jays decoy foraging crows by flying out to meet them long before they reach the vicinity of the nest. This harassing ploy seems to keep the crows concentrating their search where the jays are active, and, to my knowledge, they have yet to reach the immediate home ground of the jays.

These jays work the forest for food by hopping from branch to branch, ascending the tree to its top and then casting off to sail into another locale. Insects and nuts, acorns, and other vegetable matter provide the bulk of their diet, but they will occasionally scavenge and take the young and eggs of other birds. Our bird feeding trays are often rocked and socked by their flamboyant presence.

Scrub jay at robin's nest

Scrub Jay

Aphelocoma coerulescens

Length	11–13 in.	279.4–330.0 mm.
Wingspread	12½ in.	317.5 mm.
Weight	2¾ in.	77.9625 gm.

Before the coastal mountains around Los Angeles were flattened for homes and punctured by roadways, there were still places that youngsters could hike to and wrap up in the solitude and adventure there. Pairs of scrub jays came to scold us as we set up our camps. I remember awakening those mornings to the faint feeling of being watched and, upon looking up, I usually found a jay, head cocked over, staring me full in the face. I came home with such memories along with the usual good dose of poison oak.

This bright blue, crestless jay is represented by perhaps as many as thirteen races. With the exception of a more social Florida race (*A.c. coerulescens*) this is another jay of the West. Its range extends south from western Washington, Idaho, Utah, and Wyoming, into west Texas, the Mexican plateau, and Baja California.

The scrub jay's well-concealed nest holds from three to five eggs that are typically green with reddish-brown spots, although the California race shows some exceptional variations that may include a red background color. After the eggs have been laid in late May and June, the young are hatched in sixteen days and fledge about eighteen days later. The pair defends a territory, although the cooperative-breeding Florida race employs "helpers" in both raising the young and defending the territory.

Their diet includes seeds, nuts, insects, seasonal fruits, berries, small mammals, and the young and eggs of other birds. Like other jays, they, too, cache food for later retrieval and, like their eastern relative, the blue jay, these jays in the West have surely contributed to reforestation of oak forests when they have failed to find all of their stores or have died before they could retrieve them.

Mexican jays moving through sycamores

Mexican Jay

Aphelocoma ultramarina

Length	11½ to 13 in.	292.1 to 330.2 mm.
Wingspread	15 in.	381 mm.
Weight	3½ oz.	99.2250 gm.

The convoluted and scoured cliffs of the Chiricahua Mountains
of Arizona become haunted faces in afternoon light. In late
April fresh snows may spot the ground and one feels the cold
breath and hears whispers of old spirits. Stark sycamore arms snag
the mists and the pale-blue-backed and gray-and-white-chested
Mexican jays move in flocks over oak leaf beds, calling softly
to one another.

These jays inhabit the country southward from the lower
portions of Arizona, New Mexico, and west Texas to the heart and
western slope of Mexico. There are two races recognized.

These are among the most gregarious of jays and their communal
"extended family" nesting habits have been carefully studied.
Groups of birds assist in the construction of a single nest that
holds the three to five pale blue eggs sparingly speckled with
brown. Although it appears that only the laying female incubates
the eggs, more than a dozen flock associates assist in feeding and
protecting the young. Mexican jays do not normally begin
breeding until they are three years old.

Acorns are a staple of the Mexican jay's diet, but
probing the leaf litter they encounter and eat
a variety of insects, including weevils and beetles.
Their food tastes have also been known to
include tent caterpillars.

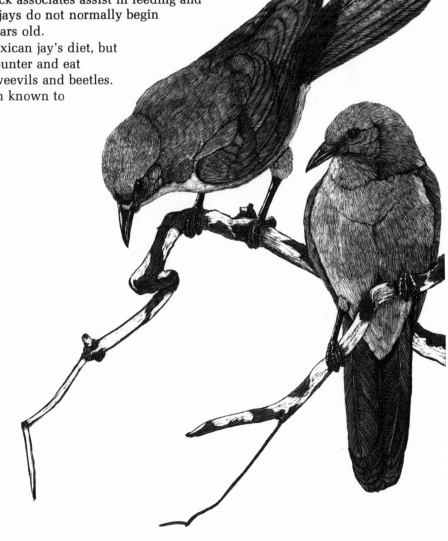

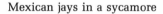
Mexican jays in a sycamore

Green jays in a thicket

Green Jay

Cyanocorax yncas

Length	11½ in.	292.1 mm.
Wingspread	15 in.	381 mm.
Weight	3½ oz.	99.225 gm.

Along the lower Rio Grande of west Texas, faded colors of river and shore melt imperceptibly into one another. Like buoyant jewels catching light, green jays bounce and dash through willows, flashing colors of blue, yellow, and green.

From southwest Texas their populations extend south through eastern Mexico and Central America to Honduras. They reappear again in the mountains of South America as far as Bolivia. It seems that this most southerly representative is distinguished by having "helpers" at the nest.

In the northern portions of their range, pairs begin nesting in May and June and have from three to five greenish-white eggs in a nest placed ten to fifteen feet above ground in heavy woods or brush.

Green jays are partial to large insects, although they will exploit opportunities to feed at picnic grounds where bits of french fries and hamburger may be found.

Studies of banded Texas jays have provided longevity records of eight and nine years for this species (Marion and Fleetwood 1974).

San Blas jays

San Blas Jay

Cissilopha san-blasiana

Length   12 in.   304.8 mm.

In late 1938 an ornithologist prowled the flat, warm desert valleys near Tucson, Arizona, seeking whatever the changing seasons might have brought in visiting bird life. Coming to a mesquite stand, he was astounded to find a family of six to eight birds that were unlike any he had ever before seen in this desert country. Among the group were two adults with rich blue backs, wings, and tails. Their underparts were solid black and their crests inconspicuous. He collected one of the adults and upon examination determined that these were San Blas jays that had somehow wandered more than seven hundred direct miles north of their normal range in Mexico. I would describe the reception of these "visitants" as tragic, for, by the end of the following year, all the San Blas jays had either been shot or had disappeared, and they were never again seen in our country.

San Blas jays normally range in western Mexico from southern Sinoloa southward to Guerrero. Sutton (1951a) is of the opinion that there are three well-defined races of this one species.

Brown jays

Plain-tailed Brown Jay

Psilorhinus morio

Length	16–18 in.	406.4–457.2 mm.
Wingspread	30 in.	762 mm.

It isn't by chance that many artists of the last twenty years who chose to portray birds carried their portfolios east to New York state to seek the criticism of Don Eckelberry. Eckelberry's portraits, like those of Fuertes before him, hold that charged vitality of the living bird as few artists have ever conveyed it. His powers of narrative, like his painting, cut directly to the essentials. There follows a description of the brown jay that, in his usual generous fashion, Don recently shared with me:

A brown jay flew across the road and we screeched to a stop. Hopping out we were in time to watch five more fly out of the scrub and land in low trees where we had a good look at them. They sounded like jays and they acted like jays, but they didn't look much like jays. They were twice the size of our familiar blue jay, as large as our common crow, and had no crests. Except for whitish underparts from chest to tail (and the legs and bill which vary with age from yellow to black) they were dull dark brown.

So far as I know, any jay has a rather labored flight, but these made it look backbreaking. No matter what the actual conditions, as I was soon to learn, they always appear to be battling a strong headwind. Down and forward go their rounded wings, down and forward, the tips curling up with the pressure, down and forward and the poor creatures have scarcely moved at all.

In the summer of 1974 the plain-tailed brown jay was rediscovered in Texas along the Rio Grande, and since that time the population has stabilized at about forty to fifty individuals (Arvin 1976). Its range extends south through portions of Nuevo Leon and Tamaulipas to Tabasco. The white-tipped brown jay, very likely another race of the brown jay, populates the regions south to Panama. In Texas, these birds form loose flocks that wander the lowland woods along the flood plain of the Rio Grande from Falcon Dam down to Fronton. In Costa Rica, Skutch (1935) observed that these jays had a number of "helpers" at their nests, and correspondence with John Arvin suggests that their northern populations in Texas may indeed have similar habits.

Their foods are of the usual jay varieties, but their larger size gives them access to prey, particularly a wide range of young birds, that are not available to smaller species. Despite their labored flight they assemble their numbers to successfully mob hawk eagles and discourage the predator from prowling their home ground.

Black-billed magpie in flight

Black-billed Magpie

Pica pica

Length	19 in.	482.6 mm.
Wingspread	24 in.	603.6 mm.
Weight	6 oz.	170.1 gm.

When magpies break from the roadside in bright, buoyant, wagtail flight, they provide an accent to the open, arid lands of the western interior. I've often gone to their haunts to sit silently beneath willow stands and listen to their songs that speak of the Columbia Plateau. They sing of bubbling, sloshing, and gurgling waters, horned owls and hawk whistles, and, sometimes, the mystery of the coyote's bark and call.

Their populations reach from Alaska south and east across the prairies of Canada and the United States to the edges of the deciduous forest. The southern limit of their range is northern Arizona, New Mexico, and, rarely, northwestern Texas.

Some four hundred vernacular names refer to the magpie, which is also a Eurasian species. Its common name is said to be a contraction of "maggot pie," a Middle English name that may have been derived from the bird's habit of plucking maggots from the backs of infected wild and domestic animals. Its more typical food includes insects, grains, and small mammals. Road kills have also been a source of food, and this taste for carrion has often proved disastrous. Feeding from poisoned carcasses set for coyotes, magpies are killed along with ravens and eagles that also take an opportunity to feed from what appears to be appropriate fare.

In spite of such travails, the magpie endures, and its nesting aggregations remain active year after year in those locales where their presence is of little competition to human enterprise.

They begin repairs on their ponderous nests as early as March, and by May six to eight greenish, brown-blotched eggs are being incubated by the female. Like others of their family, black-billed magpies are monogamous and mate for life.

Yellow-billed magpie in flight

Yellow-billed Magpie

Pica nuttalli

Length	16 in.	406.4 mm.
Wingspread	22 in.	558.8 mm.
Weight	5 oz.	141.75 gm.

As a boy, I had a yellow-billed magpie that followed me about the yard as I engaged half-heartedly in childhood duties of tidying up fallen leaves and flower beds. I think now that had it not been for my bright-colored friend, I might never have been able to build up momentum to get the tasks done. He would poise at the edge of the rake, ready to seize any grubs or crickets that my movements exposed. When his hunger had been satisfied, he remained, for he was as curious as I as to what might be revealed by the next sweep of the rake.

This species of magpie is found only in the dry, oak-clad interior valleys of northern and central California, while the black-billed occurs in California only east of the Sierra Nevada.

The yellow-billed magpie lays five to eight eggs that are spotted brown against a background of pale olive-buff. Like the black-billed, this species has domed nests that become refuges for a variety of other species, including owls, hawks, and small mammals. Its diet is also similar to that of its close relative.

As its name implies, this magpie is distinguished by its bright yellow bill. There is also an eye patch of yellow skin and a plumage that is touched with a more pronounced blue and green irridescence than that of the black-billed magpie. I have also found that the former has a more benign disposition, at least toward humans.

Ravens flying over Canyon De Chelly "White House Ruins"

Raven

Corvus corax

Length	21 in.	533.4 mm.
Wingspread	50 in.	1270.0 mm.
Weight	32 oz.	907.20 gm.

Early on a southwestern morning, I looked over the edge of
Canyon De Chelly to where it swept away below me for five
hundred feet to its dusty red floor. I followed the flights of
ebony-backed ravens that soared half that distance. They traced
the steep stone walls with wing tips and their shadows raced in
flattened and elongated pursuit over the centuries of rock. I
watched them pass over the ruins of White House, tucked back
into cool canyon recesses, and I imagined I saw faces of children
following their flights.

Audubon joined a long tradition of observers of the raven when
he saw "much in him calculated to excite our wonder." The
bird has a long history of moving humankind to comment
throughout its range in the northern hemispheres of the world. On
our continent, its range includes Alaska, Canada, the North Central
states, and south in the western United States, down through the
highlands as far as Nicaragua. In the eastern United States the bird
is abundant only in those mountains winding their way down
into Virginia and Georgia. Throughout this range there are three
races recognized, with the northern race (*C.c. principalis*) being
the largest.

Ravens lay from four to seven pale green eggs, spotted with
brown, in great, bulky nests placed in trees, cliffs, and almost any
abandoned structure that provides support. They are fond of
lining the nest with soft, decorative material that includes fur,
fresh leaves, or colored paper.

Their diets are determined by their circumstances, and they
seem capable of doing well on both fresh and scavenged foods.
In the Arctic they team up to kill young seals on the ice
floes, with one blocking retreat while another
delivers blows with its beak. When seals are not available,
they have been known to regularly consume the dung
left by dog teams. In other parts of their range they
demonstrate similar feeding versatility.

Northern ravens

White-necked ravens in courtship flight

White-necked Raven

Corvus cryptoleucus

Length	17½–21 in.	444.5–533.4 mm.
Wingspread	44 in.	1117.6 mm.
Weight	20 oz.	567.00 gms.

My family and I had traveled far into the ragged mountains of southern Arizona below Tucson. The canyons here weave freely back and forth over the border between the United States and Mexico. The cold spring mornings are dissolved by the heat of midday as the sun is high and the heat rises from the flat faces of rock. We were moving down a narrow road along Parker Canyon when we stopped at an overlook just as a pair of white-necked ravens sailed around a bend and into view. They were caught in a moment of passion and were oblivious to us, sixty feet distant. The larger male turned onto his back and sailed upside down just below and in front of the female. The lanceolate feathers of his throat glistened with oily irridescence in the strong light. He held the position for more than ten seconds, then flipped upright in a leisurely manner. In another moment the pair rounded a canyon bend and disappeared.

These ravens are normally found from southeastern Arizona and Nebraska south to Central Mexico. They share range with the common raven, but are more prone to gather in flocks. There are stories of secret reaches of Arizona and New Mexico where thousands of the birds gather each evening to roost communally.

Their five to seven eggs are green with spots or streaks of brown or purple and are placed in nests which strike bold profiles on cliff faces, telephone poles, or the odd tree that grows in their range.

Railroad beds have been picked clean by these scavengers, and studies in the 1940s suggested that the bulk of their diet was composed of insects and assorted carrion.

Trios of common crows in a pine

Common Crow

Corvus brachyrhynchos

Length	17–19 in.	431.8–482.6 mm.
Wingspread	35 in.	889 mm.
Weight	16 oz.	453.6 gm.

Along the upper Skagit River each fall and winter, runs of salmon complete their miraculous cycle, and a multitude of bird species gather to feed on their spent bodies. Traveling there many times during the winter, I draw energy from the primeval scene. A standing sea of mist is pulled from the frozen ground by the sun, and companies of crows and eagles patrol its edges. The feast brings pairs of ravens and vultures, but the crows and eagles predominate and stand feeding nearly shoulder to shoulder, showing considerable tolerance for each other in this time of plenty.

The three races of the common crow are found from the coastal reaches of Alaska and Newfoundland, south to Florida and northern Baja California.

These birds tend to nest in colonies and lay four to six brown-splotched greenish eggs some time in April or May. From their spring nesting colonies, or fall and winter roosts, they spread out over the countryside to forage for insects, grain, and carrion. In the Pacific Northwest there is little doubt that crows have adapted well to urban life, with many cities now supporting populations of these birds, along with those of starlings and pigeons.

Not long ago a careful study concluded that the northwest crow is actually a small race of the common crow (Johnston 1961), and I have chosen to follow this interpretation although the controversy has continued for over fifty years. Recently I came across some articles that the wildlife artist Allan Brooks had written early in this century, defending the "species" status for this bird. He pointed out, among other things, the bird's distinct voice and its preference for the salt water beaches of the interior waterways of British Columbia and Washington. He also noted this crow's similarity to the jackdaw. Forty years later I learned that a man had delivered what he considered the prototype of this species to the local museum. He had collected it in the Seattle watershed. The examination of this bird would have exceeded even Brooks's expectations, for it was a jackdaw. Unfortunately for this European corvid, it was an escaped captive, complete with a metal band on his leg giving the name and telephone number of his owner.

Fish crow

Fish Crow

Corvus ossifragus

Length	16–20 in.	406.4–508.0 mm.
Wingspread	32½ in.	825.5 mm.
Weight	14 oz.	396.9 gm.

Many years ago I traveled east by train in a winter of heavy snow. There was excitement in feeling warm and secure as the train plowed through drifts and the great windows framed a countryside that seemed crushed under the frigid weight of the season. When we crossed the southern reach of the Mississippi, our train stopped and several of our cars were still out on the suspension. I walked to an open window between cars and felt a ragged wind bring a flush to my face. Nothing moved, save a small contingent of crows that fed at the edge of some open water. The train lurched once, twice, and was underway again, and I walked to the end of the last car so that I might have a final glimpse of this sign of life.

Fish crows are found along beaches and their immediate waterways from Long Island south to Florida and westward to the Texas Gulf. Like their close relative, the Mexican crow, they are gregarious and maintain close proximity during all seasons, whether feeding, nesting, or roosting. Unlike the Mexican crow, fish crows soar like ravens and even hover with rapid wingbeats. Their explosive, coughing call notes are also a distinguishing characteristic.

Their nesting colonies are normally near water and often include two or three nests to a tree. Four to five greenish-blue eggs, spotted with brown, are laid in April or May.

In spring these birds traditionally work heronries for eggs and young. They are particularly adept at flying in to seize prey when the adult herons have been disturbed by humans. The bulk of their diet for the rest of the year is composed of carrion cast up by the waterways, and the insects, grains, fruits, and nuts of the adjacent lands.

Mexican crows

Mexican Crow

Corvus imparatus

Length   14–15 in.   355.6–411.0 mm.

Between and beyond cities are those repositories of waste that accumulate the spent spoils of society. Perhaps to ease our conscience, as well as to describe our methods, we now call these places "sanitary landfills," rather than dumps. Whatever we call them, they are irritating places filled with the roar of machinery that jams our discards back into the earth that provided them. Dust, dancing in air, burns the eyes and carries the stench of garbage. There is a paradox here, for, as such places repel us, they also attract other life. Amid the din and debris, crows search and feed. Whether we watch from the brow of a high canyon in the Northwest, or from a flat, dry expanse in southwest Texas, handsome black companies are out on patrol over the rubble. When some archeologist comes here from the cosmos to gather evidence to render a verdict on humankind, there will surely be crows waiting to scavenge from their camps.

Since 1968 the Mexican crow has occurred in the area around Brownsville, Texas. Its occupation of this area must be attributed, in part, to the sanitary landfill where flocks of several hundred to a thousand feed on the fish and shrimp offal dumped daily by commercial fish companies (Arvin 1976). From the Brownsville area this crow ranges west into Mexico to General Bravo and Cadereta, Nuevo Leon. Its numbers extend south to below Central Valles, San Luis Potosi, on the west, and about twenty-five miles south of Tampico in northern Veracruz on the east.

These small crows are gregarious at all periods of the year, and nest colonially between April and June, at which time their flocks retire southwestward from Brownsville to their favored nesting grounds.

A distinctive feature of this tiny crow is its frequent wing flicking. When alighting on a perch or calling, the bird opens its wings halfway and then closes them rapidly (Arvin 1976). It is further distinguished from other crows by the violet tinge of the upper body feathering and the greenish-blue of the chest and flanks. Its froglike *gar-lic* and *owwk* calls are not likely to be confused with calls of other crows.

Mexican crows are fond of hackberry and the fruit of the organ pipe cactus. Although insects too are part of their diet they are particularly fond of feeding from the carcasses of animals killed on the roadways and consuming farm produce that has fallen from trucks headed to markets.

Hawaiian crow—*alala*

Hawaiian Crow — Alala

Corvus tropicus

Length	18–20 in.	453.2–508.0 mm.
Wingspread	35 in.	889 mm.
Weight	16 oz.	453.6 gm.

In 1779 Captain Cook visited a village in what today is Kealakekua Hawaii. The people there kept tame crows about their houses; when the captain tried to secure a pair his request was denied and he was advised that the birds were sacred to the people.

Two hundred years later fewer than twenty pairs of the Hawaiian crow are to be found along the northeastern and southwestern slopes of Haualalai volcano. Unlike their mainland counterparts this species has not done well in habitat disrupted by human settlement. This aboreal corvid had evolved tastes and behaviors to match the pristine island conditions. As the land was cleared and overgrazed by cattle and pigs the crow became one of the casualties. Although the Hawaiian crows are bold and strong enough to drive off most of the predators that might reach their nests, their size is no help in sustaining their numbers in the face of a diminishing food supply. Gone is the former abundance of native fruits so favored by the *alala.* Less subtle has been a history of shooting that has apparently also reduced its numbers.

The adults hatch their grayish or greenish-blue black-flecked eggs between April and June. As the birds return to their previous nesting sites, the cutting of such habitat discourages further nest construction for the year if not longer. The blue-eyed youngsters are fed a variety of fruit, seeds, insects, and carrion. The native tree fruit (*Frey cinetia-aborea*) is an important part of their diet (Banko 1977).

Although Hawaiian crows are similar in body size to the common crow, their beaks are of a stoutness that approaches that of the raven (Captain Cook first described them as ravens — probably because of their large, ravenlike beaks). Their plumage is duller black than the common crow's, with tinges of brown in the wings. Their voice is deeper and to some more musical than that of North American crows. *Alala* means "to make a lot of noise" and in this trait they are indeed like crows everywhere.

Piñon jays engage in courtship feeding

Piñon Jay

Gymnorhinus cyanocephalus

Length	11 in.	279.4 mm.
Wingspread	18 in.	453.2 mm.
Weight	3.2 oz.	100–112 gm.

With the light still low in the east, we traveled across a plateau toward the Grand Canyon. As we crested a hill, a blue-gray cloud suddenly flushed from the sage and rose into the light like polished turquoise. These were foraging piñon jays, and they shimmered waxen as their continuous, rolling flight from ground to air, and to ground again, placed the feeding birds on new land where they could search for insects and seeds.

These colonial birds reside in the West, and breed chiefly in the piñon-juniper belt. They can be found from Oregon, Idaho, Montana, and western South Dakota, south into the mountains of California, Arizona, New Mexico, northwestern Texas, and northern Baja California.

Their four to five blue-green eggs are speckled with brown and are laid as early as late April and early May. In some parts of their range, if the piñon crop has been good, as many as three broods may be fledged by one pair in a single season.

In profile these birds look more like starlings than jays, and their walk and flight is reminiscent of a crow. There is, however, no mistaking their clamorous flocks as they forage for piñon nuts in the trees or on the ground. When such food is exhausted, the birds will, like the nutcracker, wander widely and take a variety of insects, seeds, and berries.

Clark's nutcracker on alpine fir cone

Clark's Nutcracker

Nucifraga columbiana

Length	12½ in.	306.25 mm.
Wingspread	22 in.	558.8 mm.
Weight	4 oz.	113.4 gm.

If you hike the high, rocky ridges of Mount Rainier in early spring, you'll find them silent and haunted. Clouds descend over the snow in gray, wispy waves that obscure all but the edges of distant fir stands and rock falls. Then, like some determined black and white demon, a nutcracker flies powerfully out of this mystery to pull up and land before you, testing the prospects of a handout.

Like piñon jays, Clark's nutcrackers favor the piñon-juniper belt, but the latter's range is more extensive. They prefer the high mountain country up to thirteen thousand feet, from central British Columbia to southeastern Wyoming, south into northern Baja California, Arizona, New Mexico, and northwestern Texas.

Their two to four pale green eggs are laid as early as March in an extraordinarily heavy and well-insulated nest that is nearly impossible to dislodge intact with human hands. Like their sometime mountain associate, the gray jay, Clark's nutcrackers rely on their food cache supplies to feed their young.

Capable foragers, they consume a variety of seeds and insects. When opportunity permits, they will take the eggs and young of other mountain species. They can dispatch small mammals, like chipmunks and golden-mantled ground squirrels, that are cut off from their usual retreats. In the summers they hawk moths from tree tops like flycatchers. When insect and seed supplies are low, nutcrackers will wander widely to lowland plains, and even to the Pacific Coast as far as Carmel, California.

Golden-mantled ground squirrel

Huna Raven panel

A Paradox of Myth and Culture

When we are children our curiosity is a simple formula for discovering the natural world surrounding us. As a boy I was swept up by the actions of birds. They carried my imagination to distant places and possibilities. I matched my senses with theirs and envied their powers of sight and flight. In time, I could describe their distinct features, but not without considering the common destinies we shared. I'm now dismayed to realize that there are generations without such conscious memories to link them with the earth. When our machines operate without restraint, they can cut deeply into the roots that anchor us to our earth, and spawn ethics remote from the realities of interdependence among all life forms.

Past cultures have always been moved by the beauty and strength of birds, and those of the family Corvidae have stirred a particularly wide range of responses. One member of the family, the raven, heaviest and strongest of all passerine species, has even achieved the status of a god. The raven has fired the imaginations of people around the globe, especially in the northern hemisphere. Odin, the all-father of Nordic mythology, sent a pair of ravens out at dawn to fly worldwide, and at noon they returned to perch upon his shoulders and whisper in his ears the secrets they had learned. Other Norse gods heeded his advice and Viking soldiers followed his banners into battle. Biblical writers described God-sent ravens sustaining the prophet Elijah during his retreat to the desert, and the poet Poe employed the bird as a creative focus for his maddened lament.

No tribute is more profound than that paid Raven by the Native American people of the Pacific Northwest. To the Tsimshian, Haida, Bella Bella, Tlingit, and Kwakiutl, Raven is the god who brought life and order. He is a powerful trickster who emerged from the static antedeluvian past bringing light stolen from the one who would keep the world in darkness. Raven illuminates the earth and then proceeds, with his allies and competitors, to create the essentials of fresh water, land, tides, fair weather, salmon, and even human life itself. In recognition of such power, Raven is referred to as "Real Chief" by the Haida and "Great Inventor" and "One Whose Voice Is to Be Obeyed" by the Bella Bella.

The Kwakiutl offered the afterbirth of a male newborn to ravens to peck at so that when the child was grown to manhood he would understand their cries. This interpreter could respond to nearly a dozen raven vocalizations that would tell him of a change in weather, the possibilities of attack from enemy warriors, an imminent death, or what the hunting prospects would be (Boas 1913–14). One call described was *wax, wax, wax*, which supposedly foretold a visit by a stranger. I have heard this call from the ravens near our island home, and more often than not a short time after these vocalizations someone would indeed travel up our road for a visit to us or one of our neighbors.

In a very literal sense, Raven is used to think with. His deeds explain the present order of the natural world. Humankind is placed within the cast of players that compose the theater of life. The actions of Raven provide a bridge of explanation and understanding out of the darkness of the prehuman past and into the present. In these myths there is a sense of equality between humans and other life. Man and animal spirits intermingle and become one.

The arts of the Northwest people give dramatic manifestation to the myths and form of Raven. Their sculptures in stone, wood, and bone have abstracted the essentials of "ravenness" with perfection and force. The beak, that massive and essential tool of the capricious trickster, is among the most dominant elements in their portrayals. A minimum use of primary form lines expresses eyes, wings, body, feet, and tail in a fluid form. When Raven is set in realistic silhouette, feather tracts of forehead, beak, brow, cheek, and throat are often set off in secondary form lines. Raven is portrayed in more than flat profile, and is seen in expressive action, holding sunlight, fishing, and bringing the stars and moon to humankind. The artists were sensitive to the subtle expressiveness that the living raven conveys with its feather tracts, and placed them with precision.

Haida argillite design of Raven with sun

Profile of raven with berry

In the early part of this century anthropologist Franz Boas described the elements in a Kwakiutl Raven house-front design that corresponded to the physical parts of the bird (Boas 1927). It is obvious that the strong, sweeping, abstract designs of the mythic bird are anchored in the physical reality of the actual species. The carving of the Northwest Coast reflects an even greater sensitivity to these realities. In totems, rattles, masks, and frontlets, the eyes are fashioned to angle in naturally and gaze both to the sides and forward. The mythic bird is often depicted with plumicorns erect just over the brows. In social displays of the raven bird this conveys a message of aggressive intent to other animals and it is fitting that the artist should include this expressive state in his design of the demonstrative Raven god.

The myths of Raven were a part of a social and religious tradition that was synchronized with the dominant and important elements of the Northwest Coast environment. When the new immigrants arrived, with European traditions, they were determined to impose an order of their own and had no interest in one within which Raven was featured. On the contrary, those who would follow biblical traditions had little respect for a bird that never returned to Noah when the faithful patriarch first released one in hopes that it might guide him to a landfall. It is rather ironic that in Christian literature the raven should be described leaving surviving humankind during the period of the Great Flood, for it was also during a period of awesome flooding and devastation that Raven brought light and life to the people of the Northwest. Jewish folklore tells us further that the raven had repeatedly violated the decree against love-making on the ark and was in considerable disfavor with Jehovah. When Raven flew into the lives of the native people of the Pacific Northwest, his amorous manner became a highlight of his myths.

From the beginning of European settlement in America the course of the immigrants was one of control over rather than compromise and compatibility with the existing wildlife. Ravens, crows, magpies, and jays were seen as participating culprits in crop failures, death of livestock, and diminishing populations of game animals. As omniverous feeders, the corvids did, indeed, compete for the available food supplies and often avoided the settler-farmer who attempted to destroy them. The birds were then classed as vermin, if not agents of the very devil himself. Perhaps some of these immigrants still carried memories from their heritage that stirred ancient fears of the raven, for it was the sacred raven standard that led the Viking invaders in their merciless invasions of the lands to their south. The old prayer, "Good Lord, deliver us from the raging of the Northmen," was born from engendered fears that might have been generalized to the raven as well. Edgar Allan Poe summarized his century's prevalent attitude toward the raven when he referred to the "grim, ungainly, ghastly, gaunt and ominous bird of yore." The raven was more than a calculating competitor, as his eyes had "all the seeming of a demon's that is dreaming."

By the early twentieth century, bounties of bacon, barbecues, and bullion had been offered for the destruction of corvids. The birds have been sprayed from airplanes, poisoned, and targeted by bullets and bombs. In Calaveras County, California, in the 1930s, jay hunts became social affairs, culminating in a feast and song fest hosted by the losing team from the "sportsmen's" club that killed the fewest jays. Crow roosts have been continuously bombed. One Illinois location was festooned with a thousand shrapnel grenades while the birds were out foraging. When they returned to roost, the explosives were detonated, and at daybreak one hundred thousand dead crows were on the ground. It was argued that such slaughter was necessary to protect crops and populations of livestock and game animals. Evidence to the contrary was usually ignored, as was any suggestion that corvids, to some degree, might be beneficial to human enterprise. A strange war was to be waged.

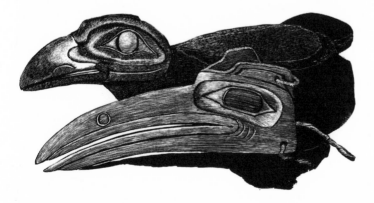

Haida stone tobacco mortar and Tlingit wooden headpiece depicting Raven

By the 1960s another refinement had been made in the disdain with which corvids were held by most people. Bert Popowski's *Varmint and Crow Hunter's Bible* describes the "hunter" shooting crows "to take up the off-season slack in available targets." Multiple kills of crows became a "chief delight," and a shooter's lifetime kill of over eight thousand crows was considered but a drop in the bucket.

Today it appears that we may be entering yet another phase of our relationships with other life in general, and corvids in particular. The Migratory Bird Treaty between the United States and Mexico was amended in 1972 to include Corvidae. As a result, some protection is afforded those of this family that were most persecuted. Scientists are finding that jays, ravens, and crows are rewarding subjects for study, as they provide new insights on animal behavior and ecology. It is the artists, however, who have come closest to moving full circle and returning to the spirit of the Raven myths. Again today, Indian and non-Indian artists alike celebrate the form and action of Raven and all his associates in nature.

Such recent activity may also suggest a desire to retrieve what Carl Jung referred to as "psychic identity." This age of diminishing species and violated environments has turned our eyes to the possibilities of our own vulnerability. We are looking about and discovering, like our "primitive" ancestors before us, that we share common destinies with other life forms. Perhaps our comparatively brief experiment with large-scale destruction of Corvidae is about to end and one of compromise is beginning. If such is the case Raven may again become a symbol for a new order and understanding.

Depictions of Raven: 1880 Bella Coola rattle, recent Kwakiutl rattle, 1880 Tlingit goat horn headpiece

Social Strategies
for Survival

Snowy owl being mobbed by crows

As a young man I made several journeys across the continent with portfolio in hand to visit those people and places that combined to form the legacy of America's artistic tribute to nature. On such a trip my wife and I came to know and become fast friends with the incomparable artist Lee Jaques and his equally distinguished wife Florence. From their friendship we derived great energy and a wealth of natural history that has influenced my interests and artistic style to this day.

One fall, fresh from such a visit with the Jaqueses, my wife and I drove north from St. Paul. We were headed to a point along the northwestern lip of Lake Superior where land forms and prevailing winds conspire to concentrate migrating raptors before they fan out over mid-continent to take up winter residency.

The hawks were there: broad-winged, red-shouldered, Cooper's, and sharp-shinned, sailing through singly and in loose groups of two's and three's. Ridges along the migration route were spotted with hawk watchers draped with cameras and scopes. We left the crowded open shoulders of the hills and worked our way into clusters of sumac to discover what accipiters might be hunting there. In the distance we could hear a high-pitched, almost frantic, squalling call. We parted leaves, and peered into the stand to find a young sharp-shinned hawk pursuing a blue jay. The hunter was leaping over branches and snatching at the tail of the breathless jay that was staying just out of reach. Then, quite suddenly, two other jays entered the stand, fresh, alert, and scolding. They bounded about and above the hawk, who turned her attention toward them. The first jay used their distraction of the hawk to make an escape to the cover of another cluster of trees. The two other jays retreated slowly into the thickest of the sumac branches leaving a fatigued hawk behind.

This pair of jays was apparently exercising an act of altruism for they were, by all appearances, risking their own survival to enhance the possible survival of another individual. By responding to the jay's distress cries and mobbing a hawk that was a threat to their own welfare, they were successful in helping the first jay escape. What seem to be altruistic acts are not uncommon in corvids. We hear stories of a wounded crow being supported and clasped by its flock associates. There is an account of a man who, discovering a crippled crow on an open sand bar, tried to dispatch the bird with a stick. The crow called frantically, summoning his brethren from surrounding trees. The flock attacked the man and forced his retreat to a distant willow thicket (Madson 1976).

Altruistic acts by corvids occur within social contexts that may enhance their survival and reproductive success. Although such behaviors also occur in other families of birds, corvids have evolved a special refinement in their use of helpers at the nest, sentries, mobbing, and flocking.

Sharp-shinned hawk pursuing blue jay

Pair of blue jays distracting hawk

Blue jay escaping

Helpers

Birds that assist a mated pair in their nest construction and cleaning, the feeding of nestlings, and post fledgling care are known as "helpers." In jays "helping" has evolved in many species. Woolfenden's studies of Florida scrub jays revealed that nesting pairs of this isolated race routinely have helpers, usually one-year-old birds that attend the nests of their parents (in later years, as new territories open up, these birds may become breeders). The helpers occasionally feed incubating females, but are most active in feeding nestlings and in assisting nest sanitation by removing fecal sacks. Once the young are fledged, helpers continue to feed them and to assist them in learning to forage and avoid predators. They are vigorous mobbers and attackers of animals intruding on the nest or young. Pairs of scrub jays having helpers have higher breeding success than pairs of birds not having helpers.

Scrub jay helpers at the nest

Mexican jay helpers at the nest

Of the New World jays, the Mexican jay has one of the most extreme forms of cooperative breeding, which might be described as an "extended family" form of the system employed by the Florida scrub jay. Each flock of Mexican jays has an exclusive home range within which there are from eight to twenty individuals. In one such flock, fourteen birds built two nests and all flock members fed the young at one of them. At the second nest eleven of the fourteen members fed the young (Brown 1974). When feeding the youngsters, helpers removed any fecal material produced. Once fledged, the young were fed by every member of the flock. Only about 26 percent of the feeding of young was by the parent birds.

Whether or not helpers are entirely altruistic is uncertain. The group may very well be providing the individual with protection against predators. Further, although scrub jay helpers do increase the reproductive success of pairs they assist, we really don't know how well these helping birds could breed on their own; for example, there may be no unoccupied territories in which to breed in a given season. Under such circumstances helping to raise brothers and sisters may result in a greater carry-over of family genes to future generations.

Piñon jay sentries

Baby Sitters and Sentries

Colonial piñon jays, too, have "helpers," but not nearly with the regularity of or on the same scale as Florida scrub and Mexican jays. There are, however, some remarkable group behaviors that work to the benefit of youngsters. In one colony of jays in northern Arizona there is a synchronization of reproductive activity. From courtship through nest building and fledging the young, all members of the flock operate within the same time frame. As a result, as soon as the young leave the nest, they are grouped in nurseries surrounded at all times by a sentry unit of adults while another group of adults is foraging. Upon the return of the foragers, the young are fed indiscriminately. The sentries, meanwhile, relieved of duty by the returning adults, fly off to gather more food. In this cycle of shared responsibilities, the young are continuously attended while a steady supply of food is provided. This protection and feeding continues for three weeks or so, after which the juveniles move with their parents to forage as a family group within the larger flock (Balda and Bateman 1972). Throughout, the young are treated solicitously by all adults of the flock, even to the extent that adults defer to them at feeding situations. Such behavior probably encourages the youngsters to remain within the social system composed of related members (Balda 1975).

After breeding season, piñon jays move as a flock, and whenever they forage there are from four to twelve sentries surrounding them. The sentry sits on an exposed or hidden high perch, and at the approach of an intruder sets up the alarm: *krawk, kraw, krawk*. In response flock members cease feeding and fly up into the trees. Sometimes they respond by flying in all directions, making it difficult for a predator to focus and capture any single individual (Balda, Bateman, and Foster 1972).

Scrub jays and crows, likewise, have their conspicuous sentinels. The latter can discriminate between humans with guns and those without. Hunters receive a quick response when spotted, and the feeding crows leave the area. Someone merely walking is scrutinized but tolerated if he keeps a reasonable distance.

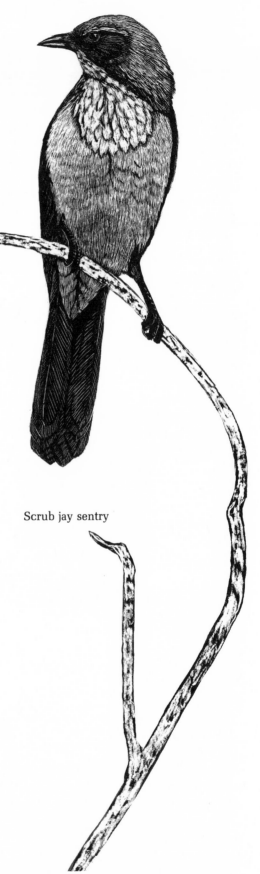

Scrub jay sentry

Mobbing

Surely one of the most conspicuous behaviors of all corvids is their mobbing. My long, silent waits in the woods for some secretive species have often been thwarted by a single jay that spotted me and summoned his associates to join him in a vigorous round of mobbing me, the intruder enemy. Mobbing is an effective social response of a group of animals to an object or individual perceived as a threat. Many corvids are particularly good at this since they already have social practices that keep their numbers in close proximity. White-necked ravens, for instance, are much more gregarious than the common raven, and when a bird of this size masses its numbers, it is something to be reckoned with. One particularly striking story was related by a fellow who had gone to visit an isolated ranch house in southern Arizona. As he approached the house a flock of some fifty ravens flew up from the scrub and began to call with ever-increasing intensity. As he moved on they circled around his head until his initial amusement gave way to a fear of attack. Some half mile down the road the bulk of the birds broke from their mobbing and flew off. He was relieved, but had a new sense of what a predator must experience when being mobbed.

The most southerly race of the green jay will quickly mob an observer that approaches and tries to locate begging young. The scolding of the adults, parents, and helpers masks the begging calls of the youngsters and makes them difficult to locate. The mobbing adult birds, moving about a predator at close quarters, are very distracting (Alvarez 1974).

Both scrub jays and Mexican jays develop an innate mobbing response to predators, which is much stronger in the former. The reason for this might be that young scrub jays are separated very early from their parents, before they have had opportunities to learn much about predators by watching the adults. Mexican jays, on the other hand, have continued association with adults of their clan and, therefore, a longer period for learning appropriate responses to predators from flock associates (Cully and Ligon 1976). Crows may also have similar extended learning periods and can be conditioned not to fear formidable predators, as some experimental captive crows associated readily with, and even begged from, a barred owl (Ramsay 1950). At our home our young ravens and magpies were at least eight weeks old before they ceased begging from yawning dogs and pant legs. Once the distinctions began to be made, however, and friends were separated from foes, any stranger entering the yard—human or otherwise—was considered a potential threat and given a thorough scolding.

Crows can be discriminating mobbers, as well as sentinels. In one locale the birds joined red-shouldered hawks in mobbing great-horned and barred owls. All the while they showed no particular fear of what we might consider to be rather fearsome associates. These same crows, however, kept their distance from, and even mobbed, red-tailed hawks. The latter was apparently just enough bigger than the red-shouldered hawk to put it into the class of an occasional crow predator (Kilham 1964).

Green jay mobbing

Crows mobbing eagle

Small flock of ravens in flight

Flocking

As a youngster I watched small flocks of crows gathering against the lemon-yellow twilight on their way to roost. Such a frolic had me convinced they were delighting in the fullness of the day past. I could see them settling into the trees and imagined them sharing the secrets that the day had brought them. On the following day pairs of birds dispersed from the flock in all directions and I was sure some had learned of a new, rich place to forage because a fellow crow had shared the news. When I later heard of a great roost in Kansas that had over five million crows, I could only sigh over what must be their collective impressions.

   Although I'm not sure that my boyhood interpretations were always correct, flocking does, indeed, provide advantages to members. It is fairly certain that a potential predator is distracted, if not confused, by a flock of birds dispersing in all directions when a predator attempts attack. Even if the flock should maintain flight formation, an avian predator such as a falcon or an accipiter might be less likely to plunge into the mass, where to do so could result in injury to itself as well as to its prey. There is, of course, always the possibility that the pursued flock can turn and take the offensive. This would be particularly true of the larger corvids who, by taking the offensive, can achieve what the solitary individual or pair of birds would never be able to do.

Flock of magpies dispersing

In foraging as well, the flock provides an advantage. The large number of birds can provide a greater likelihood of discovering and exploiting a food source that may be patchy or may come in large quantities. Their activity alone can stir up the ground litter and open up additional sources of food. The food discoveries of any one bird of the flock may benefit all. Birds of the foraging flock can also spend more time foraging and less time checking for predators as there are more pairs of eyes to do the watching. In the case of piñon jays there are a number of birds that have joined their flocks, perhaps to take advantage of the several benefits mentioned. Woodpeckers, particularly flickers, and nutcrackers and starlings are all flock associates of piñon jays (Balda, Bateman, and Foster 1972).

Routes to Sociality in Jays

The particular social arrangement through which much of a bird's behavior is expressed has been carefully studied in many of our North American jays. It may be that two alternate routes to sociality have occurred in these birds, both of which evolved from the classic territory defended by a mated pair as typified by western scrub jays (Brown 1974). One route from this point has led to the highly "colonial" organization of piñon jays. These birds are characterized by their nesting clusters, occasional helpers, and integrated flocking system. Another line of evolution involves the cooperative breeding arrangement of Florida scrub jays and Mexican jays, whose systems are distinguished by their

communal "participation of the social group in activities essential to successful reproduction and recruitment in the group" (Brown 1974). A condition intermediate between territorial and colonial exists among Steller's jays and possibly blue jays wherein a territory functions as a series of concentric zones moving outward from the immediate nesting areas. The birds' dominance diminishes as they move out from the nest (Brown 1963).

The fact that many corvids are long lived (up to twenty years for a wild crow and fourteen years for a wild blue jay) increases the likelihood that they will know each other and develop some social relationships. This is particularly true with altruistic acts as birds tend to respond altruistically to other birds of the same species when they know them.

Whatever the forces are that encourage their sociality, the results are intriguing if not captivating and the temptation to draw analogies to our own social behavior is great. Such an analogy occurred to me one early summer afternoon when I had been sitting atop a hill looking out to the house-lined street below a tiny cabin we were renting. Suddenly, with a great burst of sound and dust a delivery truck turned onto the road and narrowly missed a youngster who was playing there. A man watering his lawn at the corner registered his displeasure by blasting the truck with water as it shot by. The truck stopped abruptly and the driver got out, rolled up his sleeves, and headed for the man with the hose. As he got to the edge of the man's property he slowed down, stopped, and dared the man to step into the road. The gardener declined and continued to water his lawn. This standoff ended as half a dozen men and women emerged from

Flock of piñon jays foraging

as many homes along the street, moved to the side of their neighbor, and together confronted the "invader" truck driver. I was not so far away that I couldn't hear their raised voices scolding the delivery man. He retreated slowly to his truck and the group followed him. When he got in the truck and drove off, the unit of neighbors quickly dissolved and the individuals returned to their homes. Even before the last neighbor had disappeared I was thinking of the jays I had watched a week before. A sharp-shinned hawk had taken a pass at a fledgling jay, and a parent had challenged the hawk. Both the adult birds stood facing each other, the hawk not quite prepared to attack the emboldened Steller's jay within a few feet of its nest. In another minute four more adult jays appeared and surrounded the "invader"; this was sufficient to discourage the hawk and send him flying from the neighborhood. Although we can't describe the delivery man as a hunter like the hawk, he was nevertheless a threat to one of the members of the neighborhood below me and a potential disruption of the "peace and quiet" we generally require in the places we choose to live. In this case the human community, like the community of jays, exercised what appeared to be similar social strategies to retain the status quo. I wondered to myself if the jay behavior in this case was any less courageous than that of the people.

A Special Generalist

On an afternoon in late winter I hiked overland as a flurry of snow cut the sky in the high mountains. It was to be a short passage from trail head to shelter where we had been assured warmth against a numbing cold. I carried my one-year-old daughter beneath my coat and against my body, but the chill was penetrating and she was whimpering. I began to feel that awkward rush of vulnerability that comes when one is threatened and caught unprepared. I plodded on, head down and tottering over a trail frozen slick and solid. Rounding the edge of a snow bank, I heard a soft musical whistle sounding at once both warm and out of place in the harshness of these winter ridges. I looked up, and, a few feet off the trail, a great puffed-out gray jay was sitting on an exposed branch of alpine fir. Somehow, amid this environment so inhospitable to me, he was thriving. When we were finally inside and warmed by a fire, other jays were gathering outside the windows for handouts.

Physical Structure

Most corvids are generalists and seem to do well in marginal or disrupted habitats that would not be attractive to most other species. The gray jays I had marveled at, like others of their family, possess special physical features that appear to be critical to their survival. Certainly their loose, thick feathering provides an air-trapping insulation against the incredible cold at these altitudes. Beyond this basic equipment, the gray jay also has a very manipulative tongue and large mandibular mucus glands. Saliva secreted from these glands is used to put a sticky coat on food to be cached. The tongue can quickly roll the food into a ball, and within minutes a "bolus" is formed and placed in a suitable crevice or among needles of fir or spruce. It is these caches that sustain the adults through the winter and provide food for their young, who are often hatched when snow still tightly grips the subalpine habitat.

Nutcrackers and piñon jays possess specialized throat pouches that aid them significantly in carrying quantities of seeds to their caches. The former has a sublingual pouch that has evolved from a tongue muscle and is capable of carrying up to ninety piñon pine seeds. While carrying seeds the bird is still free to hammer, pull, and select food that it may, in the case of insects, elect to swallow. A piñon jay can carry twenty or more piñon pine seeds in its expandable esophogus (Bock, Balda, and Vander Wall 1973). Adaptations for food transportation are also found in magpies, crows, and ravens, who possess an antelingual buccal cavity for holding food. The ravens of our household would often strut from room to room seeking a proper repository for the food held in their throat. If we failed to follow them, they would regurgitate and then cache their prize in places unknown to us. A warm week later we'd be frantically searching curtain tops and undersides of cushions and couches for the source of the obnoxious smell.

Fluffy young gray jays, well insulated against the cold

Open-billed nutcracker

An essential tool of any bird is the bill, and the most striking feature of corvid bills is their generalized structure. They are sturdy and blunt enough to hammer, pry, and gape, and yet slender enough to probe. Most have the additional advantage of a slight hook to facilitate tearing, pulling, and stripping. The raven's beak is both a weapon and a tool. Since this bird often nests and forages where its neighbors may be falcons and hawks, the great bill is a necessary back-up to the raven's belligerent manner when confronting these species. It is stout enough to hold barbwire strands and bend them into superstructures for nests, as well as to deliver a *coup de grâce* to animals the size of rabbits. A great sheath of bristles protects the delicate nostrils.

The bills of the Clark's nutcracker and piñon jay are more specialized, however. Tapered and flat at the tips, they are the perfect tool for inserting into sections of cones to extract the seeds within. Such a bill is also ideal for cracking and splitting the tough shell of the fresh seed and extracting the kernel whole. The piñon jay bill is without the bristles so typical of other corvids. This enables the bird to probe more deeply in a cone without accumulating pitch over bristles. The same bill is also effective as a probe and a rake when foraging in ground litter and the substrata below (Balda and Bateman 1972). My friend Ed Sawyer, who spent most of ninety-one years watching and sketching birds, was impressed with the manner in which nutcrackers could clear or enlarge a hole by opening or gaping their bills. Once the opening had been made of sufficient size, the bird had a more appropriate location for caching or a better entry for seizing any available food within.

Gray jay on fir tree

Gray jay retrieving "bolus" for youngster

Given their large size, it is not surprising that some corvids employ their feet in a manner not unlike that of a hawk or an owl. Common crows have been observed dipping their feet into schools of herring and shad to successfully grab a meal (Hulse and Atkeson 1953). They also use their feet to seize and carry young birds and even ears of corn. In doing so they literally reach into feeding niches that may not be accessible to other passerine species. The raven's feet approach hawklike proportions and, like other raptors, the bird uses the inner toe with its heavy talons to impale its food tightly to a feeding surface, where it can be torn and stripped. Even the comparatively diminutive gray jay will use its feet to snatch and carry food, particularly that morsel that is left out and accessible in a mountain camp.

Color and Size

Being black and big also sets several corvids apart from other perching birds, and apparently works to their advantage. Blackness tends to maximize solar heat absorption. Black feathers absorb short-wave-length solar energy and decrease the temperature gradient between the skin of the bird and the outer feathers. Over a twenty-four-hour period a black bird can show a net energy advantage over a nonblack bird of similar size. This has implications for extending a bird's range into colder environments, if, indeed, blackness assists in maintenance of body temperature and conservation of body energy. It is not surprising then to find that both crows and ravens have extensive ranges that carry them into the coldest parts of the continent where they may stay year-round, to take advantage of some continuous supply of food. Crows are capable of withstanding chill Canadian winters that take temperatures as low as −40°F.

Generally speaking, northern populations are slightly larger than southern populations of birds of the same species. Northern and high-altitude corvids carry more feathering for insulation against the cold. Their size provides a greater body mass for heat production in proportion to the surface area that radiates heat.

If blackness plays any part in providing an energy advantage in more frigid habitats, it comes as a surprise when one leaves ravens over the snow banks to travel to the depths of Death Valley, only to find the same species doing quite well where afternoon temperatures may reach 120°F. "How can blackness be of any advantage here?" one asks. Perhaps even under hot desert conditions black coloring may result in energy savings. Desert mornings are cool and being black can assist in a quick warming through solar energy absorption. The raven can be done feeding before oppressive temperatures rule the day. During peak heat periods, ravens are inactive and resume foraging only with the onset of cooler afternoons and evenings. In the shade, being black is no different than being any other color, as far as thermo-regulation is concerned. These desert ravens then forage during a day's beginning and ending, hunting crepuscular prey, and scavenging in a local dump or stretch of highway.

Nutcracker with pouch full of seeds

Magpie striking ground squirrel

If being black is indeed an energy advantage, it is likely that corvids are among the birds most apt to exploit it. Not only do many of them flock and disperse to discourage and confuse a predator, but their size and belligerence would give the hunter cause for hesitation. Smaller birds without these qualities would find being black more trouble than it was worth, for they would be clear, unprotected targets for another animal seeking a meal.

The specialist species, with their foraging strategies applied to a particular habitat, will always have the edge on the more generalized feeder, so long as its particular environment or food items remain available. But this is an age of disturbed habitats, and food availabilities often fluctuate over wide ranges. New avenues for feeding are opened up for more adaptive species like most of the corvids. The suburban development of Southern California, while discouraging many passerine species, is a paradise for scrub jays. They prowl the school yards for discarded lunches and the edges of parking lots for whatever fare can be found. In the East the blue jay does equally well in and around suburban cities and homes. In the Pacific Northwest, wherever forests have been cut back and human populations have increased along with landfills and dumps, so have the crow populations. The small northwestern race of the common crow walks the city streets, front lawns, and ball fields more like an oversized starling than a corvid. Thoreau's entry about the crow from his 1859 journal is as appropriate today as it was then: "Its untamed voice is still heard above the tinkling of the forge. It sees a race pass away, but it passes not away. It remains to remind us of aboriginal nature."

Piñon jay splitting piñon nut

Raven flying with tail spread

Tool Users, Talkers, and Problem Solvers

One spring I was basking in the exuberance that occasional sunshine gives in the Northwest. Above me a crow called once, then flew from a tall fir and rose quickly in tight circles on the warm air. At perhaps two hundred feet he dropped his legs, pulled his wings straight up over his head, and parachuted a hundred feet straight down. Again he circled up on a thermal, only to backslide with tail fanned and legs dangling. For another ten minutes the crow tried a variety of half turns, walking in air, and partial slips and rolls. His acrobatics in the air were not at all unlike those I loved to perform in the sun-warmed waters of Puget Sound. My behavior was one of play as it intensified that rare feeling of buoyancy and happiness with life. I'm convinced the crow was playing as well; and although he was suspended in a different medium, he was no less delighted with life than I was.

Though it is possible that we fail to recognize it in other organisms, play seems to be exclusive to higher vertebrates. It is a mechanism of release, but more importantly it is an innocent, riskless testing and refinement of skills essential for life. Ravens have what is probably the most complex play in birds and, as Flicken (1977) points out, their intelligence permits a diversity of play which encourages the learning of relationships with their environment. This learning in turn contributes to their success in wide-ranging habitats.

Over the past ten years I've watched the play and resulting skill development in ravens that we have included in our family. In their first year the chase was a basic ingredient of their play activity, and hours were spent racing over roof tops and through the woods in hot pursuit of one another and the inconsequential item carried in the beak. One raven, Macaw by name, would strut about the yard in demonstrative display and eventually sidle up to our dozing dog and poke at his toes or tug his tail. This was the raven's invitation to chase him. On other occasions the Husky would initiate play with an unpolished rush up to the bird, a bark, and a slap of the forepaws that would come very close to pinning Macaw. Once the raven had broken into flight the frenzied chase was on, with the dog, tongue dragging, hurling himself after the bird in delight. We tended to discourage this chasing after awhile, as the participants left a wake of snapped rhododendrons and uprooted saplings.

Even in cages, ravens are fond of playing. Gwinner observed in his captive birds that one in particular was entertaining himself, and perhaps other ravens in the enclosure, by repeatedly hanging by one leg from ropes strung horizontally and performing acrobatic postures with his head and free leg (Wilson 1975). His birds, much like upright otters, were also fond of sliding down a smooth perch.

Husky chasing raven

Blue jay using tool to retrieve food

The Use of Tools

Just as play is suggestive of intelligence, the use of tools is even
more direct in demonstrating an animal's capacity to adapt to the
needs of a given situation. Captive blue jays were able to retrieve
food from outside their cage by actually manufacturing and using
a tool. Restricted by an enclosure that would not permit them to
reach food pellets with their bill alone, the birds tore out strips
of newspaper and used them as "rakes" to reach out and pull
back the food to the point where it could be retrieved and eaten
(Jones and Kamil 1973). Another captive jay raised the water
level of its drinking dish by placing solid objects in the
container. The water was thus raised to where the jay could
reach it (Judd 1975). A California scrub jay used an anvil to
crack walnuts it had stripped. Placing the nut in a pocket formed
by crossed branches, the jay then lined it up so that the seam
could be split with a series of sharp blows from the bird's beak.
All the while, the bird steadied the nut with one foot to restrict
its tendency to turn. In a month's time the jay had accumulated
over two quarts of shells beneath its anvil (Michener 1945).

   A carrion crow of northern Europe was apparently able to
utilize the same tools men were employing to retrieve a meal.
The bird watched a pair of ice fishermen haul in their lines with
fish, rebait their hooks, and return to their shelter to await more
strikes. The crow had seen the flags marking the hole in the
ice snap up, signaling a strike. After the men were settled in the
shelter, another line was struck and the small flag again flipped
up. This time, before one of the men could get out on the ice and

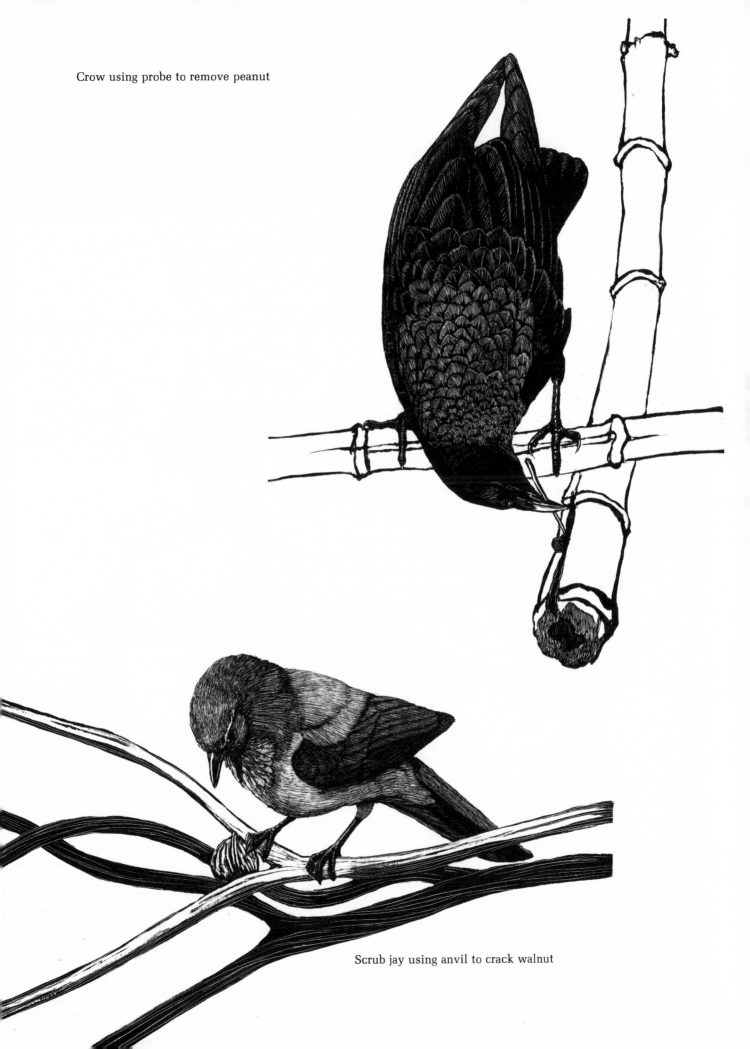

Crow using probe to remove peanut

Scrub jay using anvil to crack walnut

Crow dropping clam onto auto lanes

DEPOSIT LITTER

Crow waiting for squirrel

pull in the fish, the crow flew down, grasped the line in his beak, and pulled at it as he retreated from the hole. At a short distance he dropped the line and then walked back over it toward the hole. Keeping his weight on the line, he prevented it from slipping back into the water. Once back at the hole, he again grasped the line and retreated backwards with it. The process was repeated again and again until the fish was finally jerked onto the ice, where it was eaten.

On several occasions while waiting to catch the ferry to cross Puget Sound, I have observed a crow drop clams onto a road surface intermittently used by auto traffic. When autos were absent the crow descended and retrieved what had been crushed by the traffic. Another observer in California watched crows use road surfaces to crack walnuts by dropping them repeatedly from the air and quite possibly taking advantage of the autos that crushed them (Maple 1974).

One bird of a pair of ravens apparently used stones in defense of its nest. When climbers were descending from the nest, a rock the size of a golf ball sailed by the head of one of the men. Looking up to the edge of the cliff both men saw a bird toss one rock after another down at them—with some accuracy, as one of the climbers was struck on the leg (Janes 1976).

For several days I once observed a crow "using" a squirrel to retrieve food the bird couldn't reach in a public park refuse container. The small mammal easily pushed open the lid and prowled the refuse for food inside. When it emerged, the crow was waiting and chased the squirrel until it dropped the food. The crow retrieved the food and ate it.

Crow chasing squirrel

The Ability to Learn

Corvids have the largest cerebral hemispheres, relative to their body size, of all birds. It came as no surprise when crows, under operant conditioning methods, revealed themselves to be superior in intelligence to all other avian species tested. Among birds, their capacities for temporal discrimination were without equals (Powell 1972). Ravens, too, are remarkable learners, as our raven Macaw continually demonstrated. On one occasion I was watching him fly about the neighborhood when he suddenly plunged through a cover of trees at the end of our road. Soon he was calling riotously in a manner often associated with his frustration over trying to dislodge some object. I whistled to call him home, and he quickly appeared, flying spiritedly up the center of the road, croaking and hooting as he came. He landed and bounded over with a clothespin clasped firmly in his beak. I took it from him and rewarded him with a handful of overripe banana and chicken—his recipe for ambrosia. My generosity was to be regretted. The next morning Macaw was off with a great flourish of wing-beating to sail the length of our road and again dive into the same distant trees. Again, I heard the same impatient and raucous calling as he apparently struggled to retrieve some item. In another moment he was headed home, floating and calling in a way that can only be described as jubilant. This time, however, he was flying a white flag that streamed back from his beak. He landed and strutted over in puffed-up bravado. His prize, this time, turned out to be a pair of women's underwear.

What Macaw had learned from the day before, and I had not, was that to bring home the spoils of his neighborhood raids would result in a reward from me. It really hadn't occurred to me at the time that to give him a handful of his favorite food would have such immediate reinforcement on his retrieval behavior. I finally succeeded in extinguishing his interest in clothespins and underwear by keeping him inside on wash days.

A good deal of learning in any species will always center on food getting. A scientist friend, on a winter picnic in the Olympic Mountains of Washington state, encountered a raven where he had stopped to eat. The bird at first eagerly ate the crackers that were tossed to him. Eventually he ate his fill but was still intent on collecting the food for caching. Since picking up more than one hard cracker at a time was difficult for the bird, and at the same time he was not about to abandon the food to other foragers like gray jays, he had a problem. A solution was quickly found as the raven merely inserted each cracker edgewise into a bank of firm snow, leaving more than half the cracker exposed. Within a few minutes he had gathered a half dozen and placed them side by side in a neat, tight row; at which point he simply snapped up the bunch as a beakful and flew off to his cache. He had saved himself not only a number of separate trips, but also probably the bulk of the crackers that might otherwise have been picked up by jays in his absence.

Young ravens

J. P. Porter, early in this century, described the common crow's ability to learn by imitation. One crow mastered a novel door-opening response by watching another bird that had been trained previously to open the door in a puzzle-box experiment. The crow's good memory permitted it to apply what it had observed the other crow doing (Porter 1910).

Language and Communication

The crow's code has not been broken by any means, but we have made considerable progress toward understanding it. With its six pairs of syringial muscles the bird has the capacity to produce a range of sounds that rivals that of any other avian species. The tongue, flexible and strong, articulates these sounds into finely formed vocalizations. All corvids incorporate calls of other species to embellish their own vocabulary. Individuals may call like a child or like a loon. Whisper songs of many jays and crows include a spontaneous repertoire as diversified and lovely, by human standards, as that of any warbler, oriole, or thrush. One writer was so impressed that he wrote with great conviction of the crow's aesthetic sensitivities (Allen 1919).

Twenty-three specific crow vocalizations have been described, including those for assembly, dispersal, distress, announcement, scolding, contentment, and duet calls (Chamberlain and Cornwell 1971). Their vocalizations are most diversified and special in relation to matters of flocking and predators. The "squalling" call, for instance, is particularly effective in mustering flock members to the source of the cry. The source is usually another crow in an awkward position, quite possibly in the clutches of a predator. The caught bird has, in fact, summoned aid that comes in the form of an intensely vociferous mob of crows.

The raven of interior Alaska has a repertoire of communicative calls that is even more varied than that of the common crow. Several of their vocalizations contain two or more syllables and one, *kowulkulkulk*, contains five (Brown 1974). It is likely that the information content of the raven calls may be modified by their properties of pitch, rate, intensity, duration, and temporal patterning (the sequencing of individual calls with respect to one another). The physical displays that accompany their calls may also be modified to provide still further shading to the call's intent.

The raven's *kaaa* calls are good examples of how a vocalization may be modified and used in different contexts. The call used by a juvenile conveys an antagonistic response to a more dominant bird approaching it. Sentry adults utter an alert *kaaa* that stimulates a flock to take flight from some perceived danger. When the nesting territory is invaded, a distinct distress *kaaa* is uttered.

The chase call, *kukuk*, is often accompanied by flips to the back, barrel rolls, and even, on occasion, tumbles that carry the raven end over end through the air.

Konrad Lorenz has done as much as any man to fathom the meanings of avian communication. Some of his more provocative studies were of corvids, particularly ravens and jackdaws. When I was ten years old some thoughtful friends gave me his book, *King Solomon's Ring*, and with its reading my life's interest in birds was given new direction and maturity. Accounts of his studies are best read directly, but one of his descriptions is valuable as a preface to my own.

Lorenz's raven friend, Roah, not only communicated with the scientist with raven calls and tail wags, but also learned to use a human word in its proper context. Instead of calling *krackrackrackrack*, which is the raven's call note to "come fly," Lorenz's raven called, *roah*, with all the proper human intonation. The bird had, on numerous occasions, heard Lorenz use this raven sound as a name for the bird. In this case, however, the raven perceived the sound to be Lorenz's call note. When wishing the scientist to "come fly," or follow him, he called, *roah*, in the language of humans, not ravens.

Not too unsimilarly, our raven, Macaw, learned to say "Hello, Macaw," in the same context that I used it—as a greeting. He substituted my human greeting for his throaty *rawk*, which is given to other ravens as a greeting call. Although he occasionally resorted to his normal call note to start a day with me, as time passed he completely replaced the usual raven greeting vocalization with *hellomacaw*.

One vocalization common to a number of jays is the "whisper song." Its composition and quality are musical, and when heard by us it comes as a surprise, as we usually associate corvids with harsher sounds. The male Steller's jay seems to employ the song during courtship as it serves to reassure the female that his attentions are nonaggressive. The soft vocalization is accompanied by equally nonaggressive postures. On the ground the male sidles up to the female, singing softly, with crest lowered. The female may then respond with submissive postures, and courtship ensues.

Along our creek, jays continuously utter invitations to "come and mob." These are the *wah* calls. Over the years I have learned to distinguish between the "come mob an owl" and "come mob a squirrel." As one might expect, the former is more intense, higher pitched, and given with greater frequency. The squirrel-mobbing *wah* is rarely sustained more than a minute or two and with much less intensity. Even if heard at a distance, when subtleties are not as obvious, the owl-mobbing call is likely to be sustained for as long as it takes to drive the predator from the neighborhood, perhaps an hour or more. This type of *wah* brings in the bulk of the jays occupying the neighborhood, who are collectively emboldened. Their numbers seem to be a factor in discouraging an owl from remaining in the area. Squirrel-mobbing calls, on the other hand, seem to attract only jays of the immediate area, usually only the bird's mate. This *wah* call has its counterparts in the *weet* call of the Mexican jay and the *scree* and *whew* calls of the scrub jays.

A good deal of the jay's communicative behavior relates to the threat of predation. The *shook* call is given in flight and its basic meaning to other jays is, "Seek safety!" Jays hearing the call do just that and may fly from the ground to the cover of overhead trees or break into zigzag escape flight to the safety of the nearest ground cover. The more abrupt and harsh the *shook* call, the more likely it is to mean "hawk attack!" Like the *wah*, this call is modified in pitch and intensity to describe a particular kind of predator and, therefore, to elicit an appropriate response from those birds hearing the call. Piñon jays respond to this call by escape, and, in turn, the Steller's jays seek safety upon hearing the *krawk, kraw, krawk* warning call of the piñon jay (Balda, Bateman, and Foster 1972).

Steller's jay scolding

The *rattle* call of the Steller's jay seems to be a female vocalization exclusively, and can be given as a signal to assemble. In response to this call, both mated and unmated jays come to centers of courtship. Once assembled, the paired birds can renew bonds, and unmated birds can initiate courtship. A dominant female will also *rattle* when she supplants other jays, both male and female.

Females also utter an *ow* call which, loosely interpreted, means, "Look at me!" Along with wing fluttering and tail flickering, it serves to draw attention to the bird. When given in the company of an adult male, it may elicit courtship activity. The call and accompanying postures can also serve to convey a particular bird's dominant status.

Although Steller's jays have territories intermediate between the colonial and the single-pair-owned territory, they do defend the area immediate to their nest. The *too-leet* call can be the human equivalent to, "Hey, you're trespassing." It serves as an energy-saving mechanism. Jays hearing the vocalization often leave the vicinity of the calling jay. The territory immediate to the nest is then maintained primarily by vocal communication rather than by physical confrontation.

Distress calls are given as well. In the case of the jays I've handled, it is a high-pitched squalling call, often rasping and almost desperate in its intonation. The Steller's, scrub, and blue jays, as well as crows, almost always come to scold a marauder that holds one of their own.

Additional calls are given when birds relieve one another from nest duty. Such calls facilitate critical timing for nest relief, particularly important in view of the habitat and time of year chosen by the nutcrackers, piñon jays, and gray jays. To leave the eggs or young exposed, even for a minute, could result in their freezing. Nest-relief calls of both jays and nutcrackers are soft and secretive. The ventriloqual effect of the vocalizations makes it difficult for a predator, hearing them, to locate the nest.

Nonverbal Communication

Most corvid vocalizations are accompanied by distinct postures and feather displays. When, for instance, the Steller's jay's crest is elevated to ninety degrees, there is no mistaking that here is an aggressive bird. Tail and wing flicking communicates to other jays that the bird has dominant status and the capacity to supplant other jays. Likewise, similar elevation of the brow feathers of the common raven is expressive of dominance status and aggressive intent. The white-necked raven strikes a horizontal pose and places himself at an oblique angle to his rival. In this position his neck and upper chest feathers are elevated and their striking white basal coloring exposed. The male of this same species employs an impressive upside-down glide as part of its courtship display (see p. 36). I've observed the female of the common raven dangle her legs while hovering above the male to entice him to take flight and chase her. When corvids engage in mutual preening, physical signals are exchanged that lead to this conciliatory and pair-bonding activity.

Female raven with dangling thighs

Steller's jays with various crest elevations

White-necked raven in aggressive threat display

Green jays allopreening

Physical displays of ravens are wide ranging and, depending on the demonstrative disposition or intent of the bird, involve the elevation of head and throat feathers as well as those of chest, flanks, and rump. Submissive displays include wing and tail fanning and quivering by both sexes. Konrad Lorenz found that a highly excited male, with feathers of head and neck fully elevated, will bow low toward a female being courted and pull the nictitating membrane down over the eye. This "choking" movement demonstrates how a bird without vivid coloring can nevertheless produce impressive visual stimuli by using its available characteristics.

Our northern raven friend Macaw, like his European counterpart, Lorenz's raven Roah, was fond of giving our immediate family invitations to join him in flight. The process evolved slowly, however, and for many early weeks in his life he was content to stay with us for ground level activity. There is no question that he enjoyed the company of the dogs, infant girls, and adults, and pitched into every activity to investigate with his beak and jump back and forth from the center of things—be it picnic, gardening, or badminton game. But as his flying forays took him farther and farther away from our home, we began to notice a more concerted effort on his part to get us to join him. He would often take flight from our porch and then re-enter the open yard, coming in over the backs of our heads at good speed. Once he cleared us he would dip down into the open before us, calling, *krackrackrack*, while wagging his great black tail. The invitation is difficult to resist, even for earth-bound creatures like humans and dogs. We often spent parts of afternoons running from one end of the lawn to the other behind Macaw, who could never quite get us airborne. Dogs leaping and barking, humans laughing and running behind the bird, we'd finish the session breathless and,

Raven in flight

much to Macaw's regret, still on the ground. Such failures frustrated the raven, who would then retire to the roof to displace his anger by hammering splinters from the edges of shingles.

Our friend Macaw is now gone. His flights are surely taking him to sights and adventures that exceed anything our neighborhood provided. We remember him well and will always be deeply flattered that he should think enough of us to offer the invitation to "come fly" away to his special haunts.

Common raven in foreground strikes a submissive display

Magpie looking for food

The Energy Edge

The lives of all species, human or otherwise, are by choice or circumstance directed and fashioned by their uses and exploitation of available energy. To a large degree, life itself is a continuous process of gaining and then maintaining the level of energy necessary to sustain it. In the family Corvidae much of each species' active time is budgeted to the pursuit and maintenance of energy. A study of the yellow-billed magpie, for example, determined that the bird allocates more than 50 percent of its time to food-getting activities. Although maintenance of territory is critical for access to continuous food supply, only 3.6 percent of the bird's time is given to this activity. Being something of a colonial nester, it can share these duties with other magpie pairs. The other great block of time is given to reproduction, where the birds devote some 20 percent of their effort. The balance of their time is devoted to flying and preening (Verbeek 1972).

As in many families, it can be argued that corvids operate within the principle of "stringency" (Wilson 1975). The principle suggests that through natural selection the birds have adapted to periods of moderate or limited food availability by evolving attributes which prohibit them from making sudden responses to a superabundance of food. Most corvids, in the face of some unlimited food type, may feed and reproduce as they might when food supplies are at their normal levels. Such innate restraints, among both the more cooperative breeders or the classical territorial pairs, tend to keep their populations within the carrying capacity of their environments. As discussed below, the notable exceptions to this response are the piñon jays and nutcrackers, for in years of substantial seed supplies they will nest two or even three times in the single year.

Whatever a bird can do to enhance its access to available and dependable energy might be called an energy advantage, or edge. Certainly the success enjoyed by the corvid family can be attributed, in part, to possessing such an edge over other species. The possible energy advantages of being black and big, as is the case of crows and ravens, were explored in an earlier chapter, as were those energy benefits derived from forming colonial or communal units. Still, corvids have some other very specific behaviors and skills that enhance access to or conservation of energy.

The Nest

The three species of North American corvids that nest early in alpine and subalpine habitats build amazingly well-insulated nests, as we might expect; but their energy efficiency doesn't end here. Piñon jays, for example, place their platform of sticks on the south side of their nesting tree 85 percent of the time; that is, the nest is positioned to maximize the amount of solar energy available to incubating adults and nestlings. These birds also roost in a manner that permits taking full advantage of differential heating (Balda, Morrison, and Bement 1977). The heavy piñon jay nest platform may be composed of more than 170 sticks. If the seed harvest has been good the previous fall, birds may begin nesting as early as February when the winter sun is low. During this month, birds in most colonies place their nests out on branches to catch as much sunlight as possible. If they nest in late spring or early summer when the sun is high and intense, their structures tend to be closer to trunks where there is greater shade. From the outside to the inside, the grass lining of the nest becomes progressively finer and more shredded. Directly under the cup itself, there is a one-centimeter-thick layer of finely dissected plant parts. It is almost powdery in form and undoubtedly adds significantly to the insulation of the nest.

A pair of nutcrackers uses as many as three hundred twigs in their nest and spends approximately thirty-five hours of active time on its construction. Nesting at times as high as nine thousand feet in elevation, nutcrackers use layers of mud for the bottom of the bowl, and upon this place wood pulp. From such a base the birds form their cup of grasses and line it with soft inner bark. Cold cannot penetrate from below and the nest is so firmly placed that no storm can dislodge it.

Gray jays, also early nesters at high elevations, cache insulation material in summer and fall for later retrival during early spring of the following year. The cache provides ready access to critical materials that might otherwise be under snow at this time. The insulation is incorporated into the nest with such efficiency that the incubating jay and her eggs are kept warm and healthy at temperatures that may reach nearly −40°F (Brandt 1943).

The roofed nest of magpies breeding in hot desert areas may very well serve to control the amount of solar heat reaching both the incubating adult and the young. The nest itself is enormous and in at least one case reached seven feet in height after it had been repaired and used year after year. The pair will first gather mud or cow dung and attach this as an anchor to an appropriately large branch. Next, heavy sticks are inserted into this base in a generally upright position and, when the mud hardens, are cemented in place. A third step involves the build-up of a mud bowl within the structure, which is then lined with bark strips, feathers, horse hair, grasses, or any other soft and durable substance that will provide insulation for the eggs and young. Throughout this step-by-step construction, the exterior and roof are being completed. Construction may be carried on intermittently over a two-and-one-half-month period. An entire nest may be composed of over fifteen hundred individual branches of varying sizes.

The roofed nest also has two entrances, providing not only convenient entry and exit for the adults, but also a passage for air circulation. The roof, of course, is also a factor in thwarting predators like hawks and horned owls. Tree-climbing mammals like raccoons and foxes are not as easily deterred and account for most predation of young magpies.

Young magpies in domed nest

Recycling

Some corvids have been particularly successful in areas in which, though they may have a food supply, they do not have all the conditions necessary for reproduction. In the American Southwest, food is available but nesting trees are few and far between. In response to this, during the first half of this century the white-necked raven made some rather remarkable adaptations. Lacking the usual nest material, the ravens exploited the availability of haywire and barbed wire clippings. By traveling fence lines and fields, the ravens accumulated sufficient materials to construct the platforms and bowls of their nests almost entirely out of wire. Many of these wire nests were secured to the only elevated structures around, the cross arms of telephone poles. While the successes were admired by naturalists, who pointed to the ravens' ingenuity and efficiency, they were horrifying to the telephone company. Each spring the wire nests short-circuited the lines and cost the telephone company thousands of dollars to repair. One year the company received more than two hundred calls from users who were suffering communication problems caused by raven nests. In response, the telephone company sent out their linemen and nearly a thousand pounds of wire nests were removed from the local poles.

In eastern Washington state there is good foraging habitat for ravens, but also a dearth of trees. Here birds have nested in railroad trestles and within abandoned buildings. One pair of birds nested on the bookcase of an old school house while other birds constructed an eyrie around the moving plunging rod on a windmill. The principle materials used in its construction were wire and the rib bones of sheep. The ravens had even included a touch of decoration by including a jawbone of a cow on the front edge of the nest (Bowles and Decker 1930).

The Cache

We've spoken elsewhere of those special physical advantages that some corvids possess to aid them in their caching efforts. Certainly the pouch of the nutcracker and the esophageal cavity of the piñon jay facilitate a more efficient delivery rate, as they can carry a considerable load to the caching ground rather than making a great number of trips with single seeds. This same capacity to carry food permits piñon jays and nutcrackers to make fewer trips to the nest to feed their young than do other passerine species. When they arrive with a packed pouch or cavity, they can pump the youngsters full of food at every visit.

The caching flights of nutcrackers are often made on those warm fall days when air rises up mountain slopes. To gain elevation to cache sites in a rock bank that may be as high as 10,500 feet, the birds simply enter a thermal shell and take a free lift to the heights. Once they make their deposit, gravity brings them back down to the foraging grounds. The result is a comparatively small energy investment for a considerable energy return in the future.

Gray jay flying to cache

For at least three species of North American corvids food stored
in caches sustains the birds through the winter and provides the
energy sufficient to produce eggs and then feed the nestlings.
Gray jays, piñon jays, and nutcrackers utilize their caches for
early nesting, leaving more time for fledglings to moult, gain
weight, and develop foraging skills before the inclement weather
of the following fall and winter.

Piñon jay in profile

A cache itself can reach considerable proportions, as Balda and Vander Wall determined in their study of nutcrackers in Arizona (1976). One flock of 150 nutcrackers, living in piñon-juniper woodlands, stored nearly 300 pounds of piñon pine seeds. This represented a total of 63,000 individual seeds cached. Of that total, 130 pounds was fat, a sizable energy reserve from which to draw.

Unconsumed portions of caches often germinate and result in extension and reforestation of trees like limber, ponderosa, and piñon pine. The new trees then assist in maintenance of watersheds and soil development. These trees may also provide new food resources for these birds in the future.

The caches of both piñon jays and nutcrackers are nearly always located on the southern exposure of trees, cliffs, and open ground. These locations remain relatively free of snow and ice throughout the winter. Piñon jays are so consistent in their caching choices that one of their brethren, the Steller's jay, has learned to pilfer from these stores by always searching the south sides of trees.

Steller's jay caching nut

The life of the piñon jay is closely tied to the energy resources of the piñon nut. In some parts of its range, where the piñon crop is abundant, the piñon jay will normally nest three times a year. If, however, there is poor nut production, the late summer and early fall nesting is delayed until the following spring, as the lack of a sufficient cache will discourage earlier nesting.

Every successful species has a strategy for securing and maintaining its energy requirements. In an era of ever-increasing disrupted terrestrial environments, corvids have proved particularly adept at exploiting the available energy resources. This family may be unique in the diversity and sophistication of their strategies for gaining that elemental and essential component for viable life—the energy edge.

Inquiring brown jays

Corvids and the Community

Near the entrance to our home there is a wispy larch tree that enlivens our surroundings each spring by replacing its crusty winter attire with bristling green growth. One year a pair of bush-tits began to build their peculiarly pendulous nest into this network. It was a mistake, for though the drab nesting materials originally matched the winter larch, the nest became more and more obvious as the tree turned green. By the second week of building, a crow had found the nest, and with a few deft moves tore out the side and searched its interior. The crow's capacity to notice the out-of-the-ordinary in her territory was at its usual keen level. Only her timing was off, for the female bush-tit had yet to lay her eggs.

The crow's behavior had been more than opportunistic, for she knew her aerial and aboreal pathways well, and even subtle modifications of the tree cover could mean a nest. Crows are long lived, up to twenty years, and have time to learn the potential food sources of their communities. This crow had been resident in our locale for probably four years, and I had gotten to know her unusually distinctive scolding calls very well. It is not unlikely that she may have noticed my interest in the nesting branches of the larch, as she was fond of perching in a tall fir in our yard and surveying the activity below. My activity at the bust-tit nest may have piqued her curiosity and she investigated in my absence. In any case, the event got me wondering what effect this pair of crows, particularly the very observant female, was having on the balance of the nesting birds in our yard. The crow was successfully raiding the nests of towhees and robins, and feeding her young with their offspring. These species nested again, however, after the young crows had fledged, and this time they were successful in fledging young of their own. The kinglets, song sparrows, and Bewick wrens of our community seemed to be more successful in their initial attempts. It may well be that their smaller nests blended more cryptically with their surroundings and escaped detection by the crows. To be sure, our local family of crows later turned to other, more easily obtained fare, like emergent insects and the ever-present edible refuse from parking lots and garbage cans.

Even though corvids consume eggs and young birds, most evidence indicates that, contrary to being continuous marauders of community wildlife, corvids may actually benefit human interests through some of their food preferences. One biologist found that a single family of crows consumed more than forty thousand grubs, caterpillars, army worms, and other insects in a single nesting season (Madson 1976). It was Kalmbach (1918) who determined that nearly 20 percent of the eastern crow's diet is given exclusively to insects we consider competitive to human interests. He further concluded that more than 70 percent of the diet is vegetable in nature and, of that, only half is corn. A recent

New York state study, however, indicated that only 14 percent of the crow's diet is composed of corn. Surely there are variations from year to year, and location to location, but it does appear that many populations of crows have a distinct preference for insect larvae (Powell 1972). This preference does, indeed, contribute significantly to the control of insect populations. In any event, the crow's menu is a mixture of items we deem valuable and things we are relieved to see consumed.

When these essential roles are overlooked, the consequences can be both embarrassing and serious. A sheepman on Martha's Vineyard was convinced that local crows were solely responsible for killing his lambs. He offered a bounty of fifty cents a bird, and the crows soon disappeared, along with his money. With the crows gone, however, white grubs emerged, unchecked, and proceeded to ruin all his pastures. There was little left for his stock to feed on. With some relief, he removed the bounty, the crows returned, and the pasturage recovered (Bent 1946).

Mention has already been made of how portions of some western pine forests have been restored and extended by the unretrieved portions of piñon jay and nutcracker caches. I think, too, of the millions of acorns cached by blue jays over the centuries, which have become significant portions of the oak forests remaining today. All species interdependent with such oak forests are unknowingly indebted to the jay's habits.

Raven feeding from a road kill

The study by Turcek and Kelso (1968) of the wide-ranging corvids that store food described a number of effects that this activity has on the larger biological community. Other animals utilize the corvid's food cache to supplement their usual sources and the seed transportation and storage certainly contribute to the maintenance, renewal, and spread of tree stands. The writers also point to the fact that such plant dissemination assists in the prevention of soil erosion and exhaustion. The economic implications of all of this should be obvious as the birds are providing some savings for humankind in the costs of rearing seedlings, planting the young trees, and establishing their growth at higher elevations.

The Omnivorous Corvid

As we have noted, corvids eat, or try to eat, an amazing range of edibles, and if anything is to the bird's liking, it will surely devise a strategy to get more. A friend once watched a California scrub jay pursue and successfully catch house flies on a city lawn, while on another occasion such jays were removing ticks and deer flies from the backs and heads of mule deer. Both nutcrackers and magpies regularly remove ticks from elk and deer beds once these animals have moved out to graze.

As scavengers corvids are among the most voracious of birds. Every year ravens, crows, magpies, and jays remove thousands of pounds of carrion from highways, in the form of road kills and edible refuse. Ribbons of highways have extended their foraging habitats. We've spoken elsewhere of crow populations that have apparently expanded with the increase in garbage dumps. The birds move into areas where they are assured a suitable and usually continuous supply of food. An analysis of hundreds of castings from a large raven roost in Virginia determined that throughout the year the birds were relying principally on carrion as their source of food. Very few live animals were taken, and those that were were usually small rodents. The carrion was supplied by highways and the local dump.

In some locales the corvids' penchant for scavenging may result in their death. Animal carcasses laced with poisons, both legal and illegal, for purposes of coyote control, kill without discrimination. It's not unusual to find dead magpies, crows, and ravens at or near the site of a poisoned carcass. A poisoned bird may then become the meal of another scavenger, which in turn dies from consuming the poison contained in the body of the first.

Corvids certainly take a toll of other birds' eggs and young, but the damage they do is sometimes the result of human-created conditions. Crows can more successfully prey upon seabird and heron colonies when people flush the birds off their nests, and thereby provide an opportunity for the crows to reach the exposed eggs and young. Under normal conditions, the herons and seabirds would have remained at their nests and repelled the crows. One crow located nests by cueing in on the short canes that researchers had used to mark nest sites. Even this comparatively slight disturbance seemed to assist the crows in their predations (Picozzi 1975).

Great horned owl

As predators, corvids may, at times, have particular effects on their prey populations. Those individuals that are normally taken, however, are often in excess of what the habitat can support and may be easily caught as they are sick, injured, or enfeebled. Any predator may effectively cull out those individuals that have fewer defenses against predation and conversely encourages the development and retention of those that are resistant to predation. It is also possible that predator pressure may effect a wider distribution of the prey species. One study of carrion crows' predation habits on black-headed gulls suggested that the crows may encourage a dispersal of the nesting colony of gulls (Tinbergen 1967).

The Predator as Prey

Few animals can successfully prey to any degree on members of the corvid clan because of their size, their use of sentries, and their vigorous mobbing and flocking tendencies. Certainly, the smaller jays are subject to occasional hawk and owl predation, and even crows and magpies lose a few of their number to great horned owls, foxes, or bobcats. By and large, however, these birds enjoy a level of freedom few birds experience. They are not intimidated easily. As a matter of fact, a group of three magpies was observed successfully robbing a golden eagle of its catch. The northern raven is both tolerant of and tolerated by its large raptor associates. Along the Colville River of Alaska, ravens nest successfully in proximity to peregrine falcons, gyrfalcons, and rough-legged hawks (White and Cade 1971). Ravens in Great Britain nest near peregrines, and the better the food supply, the more tolerant are the birds of one another (Ratcliffe 1962). Ravens'

Immature peregrine falcon

Gray jays chasing weasel

territories there averaged about 6.6 square miles and were as great as 17.6 square miles. In forested conditions ravens have developed some tolerance for goshawks (Williamson and Rausch 1956). On several occasions the birds have been seen flying together without any apparent conflict. It appears that either by avoidance or by boisterous belligerence, the raven can operate successfully, even in association with some rather formidable competitors.

Magpies stealing food from eagle

Head of raven and reflecting eye

These are our North American corvids. To some, they are the apotheosis of avian form and a spirit worthy of the highest artistic tribute. Others consider them as competitors, more to be destroyed than admired. It's hard to imagine that anyone professing sensitivity would not recognize these birds as a most remarkable consolidation of highly evolved animal social systems, physical apparatus, skills, and beauty. They also demonstrate directly that often elusive capacity to sustain healthy populations within the carrying capacities of their chosen environments. To some degree, perhaps greater than most of us would admit, we find this intelligent family of birds most attractive because they are not too unlike ourselves. Their foibles are our own. They squabble within their families and wage battles with those clans that would impinge upon their home ground. Their lives involve a struggle for identity in their social hierarchy and survival in the biologic community of their choosing. Like us, they seem to have fleeting moments of joy when the mate is won, the game is played, the belly is full, and the sun shines on our backs. There is also that intriguing element about corvids that is of the unknown. These birds are more than descriptions by weight, measure, color, and distribution, for behind their amber eyes are answers to questions we may never learn to ask.

Pair of ravens over Canyon De Chelly

Bibliography

Aldous, S. E. 1942. "The White-necked Raven in Relation to Agriculture."
U.S. Fish and Wildlife Service Research Report, 5:1–56.

Allen, F. H. 1919. "The Aesthetic Sense in Birds as Illustrated by the Crow." *The Auk*, 36:112–13.

Alvarez, Humberto. 1974. "Green Jays in the Colombian Andes." Cornell University Laboratory of Ornithology, Newsletter to Members, no. 72.

Amadon, D. 1944. *The Genera of Corvidae and Their Relationships*. American Museum Novitates, no. 1251.

American Ornithologists' Union. 1957. *Check-list of North American Birds*. 5th ed. Baltimore: American Ornithologists' Union.

Angell, Tony. 1974. "The Bright, Bold Black-billed Magpie." *Pacific Search*, 9:1.

Armstrong, Edward A. 1965. *Bird Display and Behaviour*. New York: Dover.

Arvin, John. 1976. Personal correspondence with the author regarding the brown jay and Mexican crow.

———, Jimmie Arvin, Clarence Cottam, and George Unland. 1975. "Mexican Crow Invades South Texas." *The Auk*, 92:387.

Audubon, John James. 1840–44. *The Birds of America*. Vol. 4. New York: Dover Publications Reprint.

Bailey, Florence M. 1928. *Birds of New Mexico*. New Mexico Department of Game and Fish.

Balda, Russell P. 1975. Personal correspondence with the author.

Balda, Russell P., and Gary C. Bateman. 1971. "Flocking and Annual Cycle of the Piñon Jay (Gymnorhinus cyanocephalus)." *The Condor*, 73:287–302.

Balda, Russell P., and Gary C. Bateman. 1972. "The Breeding Biology of the Piñon Jay." *The Living Bird*, 11th Annual, pp. 5–42.

Balda, Russell P., Gary C. Bateman, and Gene F. Foster. 1972. "Flocking Associates of the Piñon Jay." *Wilson Bulletin*, 84:60–76.

Balda, Russell P., Michael L. Morrison, and Thomas Bement. 1977. "Roosting Behavior of the Piñon Jay in Autumn and Winter." *The Auk*, 94:494–504.

Balda, Russell P., and Stephen B. Vander Wall. 1976. "Characteristics of the Seed Catching Guild." Proceedings of the 46th Annual Meeting of the Cooper Ornithological Society, Pacific Grove, Calif.

Baldwin, S. P., and S. C. Kendeigh. 1938. "Variations in the Weight of Birds." *The Auk*, 55:415–67.

Banko, Paul. 1977. Personal discussions with the author regarding the Hawaiian crow.

Bateman, Gary C., and Russell P. Balda. 1973. "Growth, Development, and Food Habits of Young Piñon Jays." *The Auk*, 90:39–61.

Baumel, Julian J. 1953. "Individual Variation in the White-necked Raven." *The Condor*, 55:26–32.

———. 1957. "Individual Variation in the Fish Crow (Corvus ossifragus)." *The Auk*, 74:73–78.

Beidleman, R. B., and J. H. Enderson. 1964. "Starling-Piñon Jay Associations in Southern Colorado." *The Condor*, 66:437.

Bent, A. D. 1946. *Life Histories of North American Jays, Crows, and Titmice*. U.S. National Museum Bulletin, no. 191.

Blackburn, Carol Finley. 1968. "Yellow-billed Magpie Drowns Its Prey." *The Condor*, 70:281.

Boas, Franz. 1913–14. *35th Annual Report of the Bureau of American Ethnology*, part 1, p. 606.

———. 1927. *Primitive Art*. New York: Dover Publications Reprint, 1955.

Bock, W. J. 1961. "Salivary Glands in the Gray Jays (Perisoreus)." *The Auk*, 78:355–65.

———, Russell P. Balda, and Stephen B. Vander Wall. 1973. "Morphology of the Sublingual Pouch and Tongue Musculature in Clark's Nutcracker." *The Auk*, 90:491–519.

Bowles, J. H. 1900. "The Northwest Crow." *The Condor*, 2:84–85.

———, and F. R. Decker. 1930. "The Ravens of the State of Washington." *The Condor*, 32:192–201.

Braly, J. C. 1931. "Nesting of the Piñon Jay in Oregon." *The Condor*, 33:29.

Brandt, Herbert. 1943. *Alaska Bird Trails*. Cleveland, Ohio: Bird Research Foundation.

———. 1951. *Arizona and Its Bird Life*. Cleveland, Ohio: Bird Research Foundation.

Brooks, Allan, 1942. "The Status of the Northwestern Crow." *The Condor*, 44:166.

Brown, J. L. 1963. "Aggressiveness, Dominance, and Social Organization in the Steller's Jay." *The Condor*, 65:460–84.
———. 1964. *The Integration of Agonistic Behavior in the Steller's Jay (Cyanocitta stelleri) (Gmelin)*. Berkeley: University of California Press.
———. 1970. "Cooperative Breeding and Altruistic Behaviour in the Mexican Jay (Aphelocoma ultramarina)." *Animal Behaviour*, 18:366–78.
———. 1972. "Communal Feeding of Nestlings in the Mexican Jay (Aphelocoma ultramarina): Interflock Comparisons." *Animal Behaviour*, 20:395–403.
———. 1974. "Alternate Routes to Sociality in Jays—with a Theory for the Evolution of Altruism and Communal Breeding." *American Zoologist*, 14:63–80.
Brown, Roderick Neil. 1974. "Aspects of Vocal Behavior of the Raven (Corvus corax) in Interior Alaska." Thesis, University of Alaska, Fairbanks.

Carpenter, Edmund. 1976. "Art of the Northwest Coast Indians: Collectors and Collections." *Natural History*, 85:56–67.
Chamberlain, D. R., W. B. Gross, G. W. Cornwell, and H. S. Mosby. 1968. "Syringeal Anatomy in the Common Crow." *The Auk*, 85:244–52.
Chamberlain, Dwight R., and George W. Cornwell. 1971. "Selected Vocalization of the Common Crow." *The Auk*, 88:613–34.
Chapman, Frank. 1908. *Camps and Cruises of an Ornithologist*. New York: D. Appleton and Co.
Coe, Ralph T. 1976. *Sacred Circles: Two Thousand Years of North American Indian Art*. Seattle: University of Washington Press.
Cottam, Clarence. 1945. "Feeding Habits of the Clark Nutcracker." *The Condor*, 47:168.
Crossin, R. S. 1967. "The Breeding Biology of the Tufted Jay." *Proceedings of the Foundation of Vertebrate Zoology*, 1:265–300.
Cully, Jack F. R., and J. David Ligon. 1976. "Comparative Mobbing Behavior of Scrub and Mexican Jays." *The Auk*, 93:116–25.

Davis, John. 1951. "Notes on the Nomenclature of the Brown Jays." *The Condor*, 53:152–53.
———, and Laidlaw Williams. 1957. "Irruptions of the Clark Nutcracker in California." *The Condor*, 59:297–307.
Davis, L. Irby. 1958. "Acoustic Evidence of Relationship in North American Crows." *Wilson Bulletin*, 70:151–67.
———. 1972. *Birds of Mexico and Central America*. Austin and London: University of Texas Press.
Devitt, O. E. 1961. "An Example of the Whisper Song of the Gray Jay (Perisoreus canadensis)." *The Auk*, 73:265.
Dixon, J. B. 1956. "Clark Nutcrackers Preying on Ground Squirrels and Chipmunks." *The Condor*, 58:386.
Dixon, Joseph S. 1933. "Three Magpies Rob a Golden Eagle." *The Condor*, 35:161.
———. 1944. "California Jay Picks Ticks from Mule Deer." *The Condor*, 46:204.
Dow, Douglas D. 1965. "The Role of Saliva in Food Storage by the Gray Jay." *The Auk*, 82:139–54.

Emlen, J. T., Jr. 1936. "Age Determination in the American Crow." *The Condor*, 38:99–102.
———. 1940. "The Midwinter Distribution of the Crow in California." *The Condor*, 42:287–94.
———. 1942. "Notes on a Nesting Colony of Western Crows." *Bird Banding*, 13:143–54.
Erickson, Mary M. 1937. "A Jay Shoot in California." *The Condor*, 39:111–14.
Erpino, Michael J. 1968a. "Aspects of Thyroid Histology in Black-billed Magpies." *The Auk*, 85:397–403.
———. 1968b. "Nest-related Activities of Black-billed Magpies." *The Condor*, 70:154–65.
———. 1969. "Seasonal Cycle of Reproductive Physiology in the Black-billed Magpie." *The Condor*, 71:267–79.
Evenden, Fred G., Jr. 1947. "Nesting Studies of the Black-billed Magpie in Southern Idaho." *The Auk*, 64:260–65.

Flicken, Millicent. 1977. "Avian Play." *The Auk*, 94:573–82.
Forbush, E. H. 1927. *Birds of Massachusetts and Other New England States*. Vol. 2. Norwood, Mass.: Department of Agriculture.
French, N. R. 1955. "Foraging Behavior and Predation by Clark Nutcracker." *The Condor*, 57:61–62.
Frings, Hubert, and Mabel Frings. 1957. "Recorded Calls of the Eastern Crow as Attractants and Repellents." *Journal of Wildlife Management*, 21:91.
———. 1959. "The Language of Crows." *Scientific American*, 201:119–31.
———. 1964. *Animal Communication*. New York: Blaisdel Publishing Co.

Geist, Otto W. 1936. "Notes on a Fight between Alaska Jays and a Weasel." *The Condor*, 38:174.

Goodwin, Derek. 1976. *Crows of the World.* Ithaca, N.Y.: Cornell University Press.

Goulden, Loran L. 1975. "Magpie Kills a Ground Squirrel." *The Auk*, 92:606.

Griffin, Homer V. 1952. "Porcupine Quill Fatal to Jay (Steller's)." *The Condor*, 54:364.

Grimes, S. A. 1940. "Scrub Jay Reminiscences." *Bird Lore*, 42:431–36.

Gross, A. O. 1949. "Nesting of the Mexican Jay in the Santa Rita Mountains, Arizona." *The Condor*, 51:241–49.

Haderlie, Eugene C., and Aileen E. Haderlie. 1959. "Longevity of an Injured Scrub Jay." *The Condor*, 61:60.

Hamilton, W. D. 1963. "The Evolution of Altruistic Behavior." *American Naturalist*, 97:354–56.

Hardy, J. W. 1961. "Studies in Behavior and Phylogeny of Certain New World Jays (Garrulinae)." *University of Kansas Science Bulletin*, 42:13–149.

———, 1969. "A Taxonomic Revision of the New World Jays." *The Condor*, 71:360–75.

———, and Ralph J. Raitt. 1976. "There Are Two Distinctive Races of San Blas Jays not in Geographic Contact." Proceedings of the 46th Annual Meeting of the Cooper Ornithological Society, Pacific Grove, Calif.

Harlow, Richard F., Robert G. Hooper, Dwight R. Chamberlain, and Hewlette S. Crawford. 1975. "Some Winter and Nesting Season Foods of the Common Raven in Virginia." *The Auk*, 92:298–306.

Heppner, Frank. 1970. "The Metabolic Significance of Differential Absorption of Radiant Energy by Black and White Birds." *The Condor*, 72:50–59.

Hering, Paul E. 1934. "The Food of the American Crow in Central New York State." *The Auk*, 51:470–76.

Holm, Bill. 1965. *Northwest Coast Indian Art: An Analysis of Form.* Seattle and London: University of Washington Press.

Hooper, Emmet T. 1938. "Another Jay Shoot in California." *The Condor*, 40:162–63.

Hough, John N. 1949. "Steller's Jay Flies South in the Spring." *The Condor*, 51:188–89.

Hubbard, John P., and David M. Niles. 1975. "Two Specimen Records of the Brown Jay from Southern Texas." *The Auk*, 92:797.

Hulse, David C., and Thomas Z. Atkeson. 1953. "Fishing by the Common Crow." *The Auk*, 70:373.

Hunter, Maxwell W., and Alan C. Kamil. 1975. "Marginal Learning-set Formation by the Crow (Corvus brachyrhynchos)." *Bulletin of the Psychonomic Society*, 5:373–75.

Ivor, H. R. 1958. "Steller's Jay Anting with Tobacco Smoke." *Wilson Bulletin*, 70:288.

Janes, Stewart. 1976. "The Apparent Use of Rocks by a Raven in Nest Defense." *The Condor*, 78:409–23.

Jewett, Stanley G. 1924. "An Intelligent Crow." *The Condor*, 26:72.

Johnson, H. C. 1902. "The Pinyon Jay." *The Condor*, 4:14.

Johnston, D. W. 1961. *The Biosystematics of American Crows.* Seattle: University of Washington Press.

Johnston, Richard F. 1958. "Function of Cryptic White in the White-necked Raven." *The Auk*, 75:350–51.

Jones, T. B., and A. C. Kamil. 1973. "Toolmaking and Tool-using in the Northern Blue Jay." *Science*, 180:1076–78.

Judd, W. W. 1975. "A Blue Jay in Captivity for Eighteen Years." *Bird Banding*, 46:250.

Kalmbach, E. R. 1918. *The Crow and Its Relation to Man.* U.S. Department of Agriculture Bulletin, no. 621.

Kendeigh, Charles S. 1969. "Energy Responses of Birds to Their Thermal Environments." *Wilson Bulletin*, 81:441–49.

———. 1970. "Energy Requirements for Existence in Relation to Size of Bird." *The Condor*, 72:60–65.

Kennard, J. H. 1975. "Longevity Records of North American Birds." *Bird Banding*, 46:55–73.

Kilham, Lawrence. 1960. "Eating of Sand by Blue Jays." *The Condor*, 62:295.

———. 1964. "Interspecific Relations of Crows and Red-shouldered Hawks in Mobbing Behavior." *The Auk*, 66:247.

Lawrence, Louise de Kiriline. 1957. "Displacement Singing in a Canada Jay (Perisoreus canadensis)." *The Auk*, 74:260–61.

———. 1968. "Notes on Hoarding Nesting Material, Display, and Flycatching in the Gray Jay (Perisoreus canadensis)." *The Auk*, 85:139.

Lepthien, Larry W. 1976. "Changing Winter Distribution and Abundance of the Blue Jay." Proceedings of the 46th Annual Meeting of the Cooper Ornithological Society, Pacific Grove, Calif.

Ligon, J. D. 1971. "Late Summer-Autumnal Breeding of the Piñon Jay in New Mexico." *The Condor*, 73:147–53.

Lorenz, K. Z. 1952. *King Solomon's Ring: New Light on Animal Ways.* New York: Thomas Y. Crowell Co.

———. 1968. "Pair Formation in Ravens." In *Man and Animal: Studies in Behavior*, ed. Heinz Friedrich. New York: St. Martin's Press.

Lucid, V. J., and R. N. Conner. 1974. "A Communal Common Raven Roost in Virginia." *Wilson Bulletin*, 86:82–83.

Madson, John. 1976. "The Dance on Monkey Mountain." *Audubon Magazine*, 78:52–61.

Malin, Edward, and Norman Feder. 1968. *Indian Art of the Northwest Coast.* Denver Art Museum.

Maple, Terry. 1974. "Do Crows Use Automobiles as Nutcrackers?" *Western Birds*, 5:97–98.

Marion, W. R., and R. J. Fleetwood. 1974. "Longevity of Green Jays." *Bird Banding*, 45:178.

Maser, Chris. 1975. "Predation by Common Ravens on Feral Rock Doves." *Wilson Bulletin*, 87:552–53.

Mewaldt, R. L. 1952. "The Incubation Patch of the Clark Nutcracker." *The Condor*, 54:361.

———. 1956. "Nesting Behavior of the Clark Nutcracker." *The Condor*, 58:3–23.

———. 1958. "Pterylography and Natural and Experimentally Induced Molt in Clark's Nutcracker." *The Condor*, 60:165–87.

Michener, Josephine R. 1945. "California Jays, Their Storage and Recovery of Food and Observations at One Nest." *The Condor*, 47:206–10.

Miller, F. W. 1952. "Blue Jay (Cyanocitta cristata), 'Anting' with Burning Cigarettes." *The Auk*, 69:87–88.

Mugaas, John N., and James R. King. 1976. "Behavioral Energetics of Free-living Black-billed Magpies (Pica pica hudsonia), near Pullman, Wash." Proceedings of the 46th Annual Meeting of the Cooper Ornithological Society, Pacific Grove, Calif.

Murray, J. J. 1949. "Nesting Habits of the Raven in Rockbridge County, Va." *Raven*, 20:40–43.

Nelson, A. L. 1934. "Some Early Summer Food Preferences of the American Raven in Southeastern Oregon." *The Condor*, 36:10–15.

Orenstein, Ronald I. 1972. "Tool-Use by the New Caledonian Crow (Corvus monaduloides)." *The Auk*, 89:674.

Phillips, Allan R. 1950. "The San Blas Jay in Arizona." *The Condor*, 52:86.

———, J. Marshall, and G. Monson. 1964. *The Birds of Arizona.* Tucson: University of Arizona Press.

Picozzi, N. "Crow Predation on Marked Nests." *Journal of Wildlife Management*, 39:151–55.

Pitelka, Frank A. 1945a. "Differentiation of the Scrub Jay (Aphelocoma coerulescens) in the Great Basin and Arizona." *The Condor*, 47:23–26.

———. 1945b. "Pterylography, Molt, and Age Determination of American Jays of the Genus 'Aphelocoma.'" *The Condor*, 47:229–61.

———. 1951. "Speciation and Ecological Distribution in American Jays of the Genus 'Aphelocoma.'" *University of California Publications in Zoology*, 50:195–465.

———. 1958. "Timing of Molt in Steller's Jays of the Queen Charlotte Islands, British Columbia." *The Condor*, 60:38–49.

Popowski, Bert. 1962. *The Varmint and Crow Hunter's Bible.* New York: Doubleday and Co.

Porter, J. P. 1910. "Intelligence and Imitation in Birds: A Criterion of Imitation." *American Journal of Psychology*, 21:1–71.

Potter, E. F. 1970. "Anting in Wild Birds: Its Frequency and Probable Purpose." *The Auk*, 87:692–713.

———, and Doris C. Hauser. 1974. "Relationship of Anting and Sunbathing to Molting in Wild Birds." *The Auk*, 91:537–63.

Powell, Robert W. 1972. "Operant Conditioning in the Common Crow (Corvus brachyrhynchos)." *The Auk*, 89:738–42.

————. 1973. "Time-based Responding in Pigeons and Crows." *The Auk*, 90:803–8.

————. 1974a. "Comparison of Differential Reinforcement of Low Rates (DRL) Performance in Pigeons (Columba livia) and Crows (Corvus brachyrhynchos)." *Journal of Comparative and Physiological Psychology*, 4:736–46.

————. 1974b. "Some Measures of Feeding Behavior in Captive Common Crows." *The Auk*, 91:571–74.

Ramsay, A. O. 1950. "Conditioned Responses in Crows." *The Auk*, 67:456–59.

Ratcliffe, D. A. 1962. "Breeding Density in the Peregrine (Falco peregrinus) and Raven (Corvus corax)." *Ibis*, 104:13–39.

Reid, William. 1976. "The Raven Carving." Audiovisual script, Museum of Anthropology, University of British Columbia, Vancouver.

Richardson, Frank. 1938. "California Jays Catch Flies." *The Condor*, 40:264.

Rockwell, R. B. 1909. "The Use of Magpies' Nests by Other Birds." *The Condor*, 11:90–92.

Roth, Vincent D. 1970. "Unusual Predatory Activities of Mexican Jays and Brown-headed Cowbirds under Conditions of Deep Snow in Southeastern Arizona." *The Condor*, 73:113.

Sawyer, E. J. 1925. "Rocky Mountain Jay Using Its Feet for Carrying Purposes." *The Condor*, 27:36.

Schorger, A. W. 1941. "The Crow and the Raven in Early Wisconsin." *Wilson Bulletin*, 53:103–6.

Schwan, Mark W., and Darrell D. Williams. 1976. "Temperature Regulation in the Common Raven of Interior Alaska." Proceedings of the 46th Annual Meeting of the Cooper Ornithological Society, Pacific Grove, Calif.

Scott, Jack Denton. 1974. "Woe to the Farmer's Foe, the Crow." *National Wildlife*, 12:44–47.

Selander, Robert K. 1959. "Polymorphism in Mexican Brown Jays." *The Auk*, 76:385–417.

Shifflett, Wayne A. 1975. "First Photographic Record of the Brown Jay in the United States." *The Auk*, 92:797.

Shufeldt, R. W. 1890. *The Myology of the Raven*. London and New York: Macmillan and Co.

Skinner, M. P. 1921. "Notes on the Rocky Mountain Jay in the Yellowstone National Park." *The Condor*, 23:147–51.

Skutch, A. F. 1935. "Helpers at the Nest." *The Auk*, 60:257–73.

————. 1976. *Parent Birds and Their Young*. Austin and London: University of Texas Press.

Smith, R. Bosworth. 1909. *Bird Life and Bird Lore*. London: John Murray.

Stettenheim, Peter R. 1974. "The Bristles of Birds." *The Living Bird*, 12:201–34.

Stone, Peter. 1970. "Reciprocity: The Gift of a Trickster." Paper read for symposium, "The Social Use of Metaphor." 69th Annual Meeting, American Anthropological Association, San Diego, Calif.

Sutton, George Miksch. 1951a. *Mexican Birds: First Impressions*. Norman: University of Oklahoma Press.

————. 1951b. "Subspecific Status of the Green Jays of Northeastern Mexico and Southern Texas." *The Condor*, 53:124–28.

————. 1972. *At a Bend in a Mexican River*. New York: Paul S. Ericksson.

————. 1975. *Portraits of Mexican Birds*. Norman: University of Oklahoma Press.

————, and Perry W. Gilbert. 1942. "The Brown Jay's Furcular Pouch." *The Condor*, 44:160–65.

Thompson, N. S. 1969. "Caws and Affect in the Communication of Common Crows." *Bulletin of the Ecological Society of America*, 50:142.

Thompson, W. A., I. Vertinsky, and J. R. Krebs. 1974. "The Survival Value of Flocking in Birds: A Simulation Model." *Journal of Animal Ecology*, 43:785–820.

Thorp, W. A. 1963. *Learning and Instinct in Animals*. Cambridge, Mass.: Harvard University Press.

Tinbergen, N., M. Impekoven, and D. Franck. 1967. "An Experiment on Spacing-Out As a Defence against Predation." *Behavior*, 28:307–21.

Todd, Kenneth S., Jr. 1968. "Weights of Black-billed Magpies from Southwestern Montana." *The Auk*, 58:508.

Turcek, F. J., and L. Kelso. 1968. "Ecological Aspects of Food Transportation and Storage in the Corvidae." *Communications in Behavioral Biology*, 1:277–97.

Tyrrell, W. Bryant. 1945. "A Study of the Northern Raven." *The Auk*, 62:1–7.

Mobbing crow attacked by merlin

Verbeek, Nicholaas A. M. 1972. "Daily and Annual Time Budget of the
Yellow-billed Magpie." *The Auk*, 89:567–82.
———. 1973. "The Exploitation System of the Yellow-billed Magpie."
University of California Publications in Zoology, 99:1–58.

Walsh, Francis J. 1976. "Lyndon Is for the Birds, Just so They're Crow."
Smithsonian, 7:204.
Warren, O. B. 1899. "A Chapter in the Life of the Canada Jay." *The Auk*,
16:12–19.
Weisbrod, A. R. 1971. "Grooming Behaviors of the Blue Jay." *The Living Bird*,
10:271–84.
Westcott, Peter W. 1964. "Invasion of Clark Nutcrackers and Piñon Jays into
Southeastern Arizona." *The Condor*, 66:441.
———. 1969. "Relationships among Three Species of Jays Wintering in
Southeastern Arizona." *The Condor*, 71:353–59.
White, C. M., and T. J. Cade. 1971. "Cliff-nesting Raptors and Ravens along the
Colville River in Arctic Alaska." *The Living Bird*, 10:107–50.
Wiggins, Ira L. 1947. "Yellow-billed Magpies' Reaction to Poison." *The Condor*,
49:213.
Williamson, Francis S. L., and Robert Rausch. 1956. "Interspecific Relations
between Goshawks and Ravens." *The Condor*, 58:165.
Wilson, Edward O. 1975. *Sociobiology: The New Synthesis.* Cambridge, Mass.:
Harvard University Press.
Woolfenden, Glen E. 1973. "Nesting and Survival in a Population of Florida
Scrub Jays." *The Living Bird*, 12:25–49.
———. 1975. "Florida Scrub Jay Helpers at the Nest." *The Auk*, 92:1–15.
———, and John W. Fitzpatrick. 1976. "Intra-familial Dominance in the Florida
Scrub Jay." Proceedings of the 46th Annual Meeting of the Cooper Ornithological
Society, Pacific Grove, Calif.